# REEFS OF TIME

LISA S. GARDINER

# REEFS OF TIME

**WHAT FOSSILS REVEAL ABOUT CORAL SURVIVAL**

PRINCETON UNIVERSITY PRESS
PRINCETON & OXFORD

Copyright © 2025 by Lisa S. Gardiner

Princeton University Press is committed to the protection of copyright and the intellectual property our authors entrust to us. Copyright promotes the progress and integrity of knowledge created by humans. Thank you for supporting free speech and the global exchange of ideas by purchasing an authorized edition of this book. If you wish to reproduce or distribute any part of it in any form, please obtain permission.

Requests for permission to reproduce material from this work should be sent to permissions@press.princeton.edu

Published by Princeton University Press
41 William Street, Princeton, New Jersey 08540
99 Banbury Road, Oxford OX2 6JX

press.princeton.edu

All Rights Reserved

ISBN 978-0-691-24733-5
ISBN (e-book) 978-0-691-24731-1

British Library Cataloging-in-Publication Data is available

Editorial: Alison Kalett and Hallie Schaeffer
Production Editorial: Kathleen Cioffi
Text and Jacket Design: Heather Hansen
Production: Jacquie Poirier
Publicity: Matthew Taylor and Kate Farquhar-Thomson
Copyeditor: Laurel Anderton

Jacket images: Historic Illustrations / Alamy Stock Photo

This book has been composed in Arno with Franklin Gothic URW and Nave

Printed in the United States of America

10 9 8 7 6 5 4 3 2 1

*To Al Curran, who introduced me, and many others, to fossil reefs*

## CONTENTS

|    | *Introduction*                                              | 1   |
|----|-------------------------------------------------------------|-----|
| 1  | Welcome to the Rock Factory                                 | 12  |
| 2  | The Present as the Key to the Past                          | 32  |
| 3  | Reefs at the Shallow End of Deep Time                       | 54  |
| 4  | Into the Death Assemblage                                   | 75  |
| 5  | When Reefs Fall Apart                                       | 93  |
| 6  | When Reefs Persist                                          | 121 |
| 7  | Lying Low to Avoid Extinction                               | 140 |
| 8  | Designing Reefs That Can Survive Us                         | 168 |
| 9  | Survival of the Heat Tolerant                               | 193 |
| 10 | The Anthropocene Coral Paradox and the Future of Reefs      | 214 |
|    | *Acknowledgments*                                           | 235 |
|    | *Notes*                                                     | 237 |
|    | *Index*                                                     | 267 |

# REEFS OF TIME

# INTRODUCTION

An urban park bench is about the last place I would expect to find a fossil reef, but there's no mistaking the corals within the stone benches at Miami's Maurice A. Ferré Park. Instead of sitting on a bench, I'm stooped over inspecting one as joggers sprint past, music blares from somewhere, and kids shriek while playing in the grass nearby. As I look more closely at the large rectangle of stone that forms the seat of the bench, I can see the shapes of star coral skeletons, pockmarked with tiny stars where the coral animals once lived. And there are some with larger star shapes, called great star coral. The sinuous lines of brain coral wind across the left end of the park bench, and numerous fossil clam shells are suspended in the rock on the right side. The curved shells look like smiles in cross section. I stand up to find a dog walker looking at me as if I'm crazy. His suspicions are probably confirmed when I start taking photos of the bench. But I can't help myself—the fossil reef within it looks so much like the fossil reefs I used to study. This limestone rock, formed from a coral reef long ago, was excavated from a quarry and then made into benches for an oceanfront park in downtown Miami with a backdrop of skyscrapers, but it's still a fossil reef to me. It's a piece of coral reef history.

If you were to sit on the bench's fossil reef and look out to sea, you would get a view of the Port of Miami. A lineup of massive cruise ships takes up one side of the paved, human-built island at the port's center. On the island's other side are container ships and giant metal cranes that look somewhat like the AT-AT walkers from *Star Wars*. Despite the entirely unnatural concrete and steel environment of the port, and the potential dangers of pollution, extreme heat, and other human-caused problems, small corals live below the lines of ships. A few months before I found the fossil corals in this bench, researchers published a study about the living corals that are able to persist in the nonideal waters of the port.[1] There's other life in the port too. Coral City Camera, an underwater live stream from the seafloor below the port, regularly captures images of fish, rays, and dolphins as they swim by.[2] But not all marine life can survive the port. And even in parts of the ocean that are arguably far more natural than the port, it's become more challenging for tropical marine life to survive. Overall, the ocean is becoming a less hospitable place.

A stone's throw from the park bench and the port, other corals live in the Frost Museum of Science. There, researchers are growing small coral colonies in saltwater tanks to help species survive. Some of the new colonies are on display in an exhibit with blue-tinted light, each growing on its own tiny pedestal. Looking through the glass of the tank at the corals organized in rows is like looking through the glass wall of a hospital nursery to see the newborns. When I visited, staghorn coral and finger coral were growing in the nursery. If all goes well, one day those corals will be thriving in the sea.

The corals in the park bench lived in a past world without much, if any, human influence. The corals currently living in the Port of Miami are somehow persisting today despite our environmental mess. And the corals in the Frost Museum of Science will be used in the future to rebuild reefs we've destroyed. After spending more than a year writing about the past, present, and possible future of coral reefs, I felt it was apropos to find all three together in the shadow of the Miami skyline. All three are related—the stories of how reefs lived in the past, when they thrived and when they floundered, can help us understand reefs today and find ways to help reefs persist into the future, as we'll explore in this book.

The living corals in the port and in the museum, like those that we see snorkeling, are typically colonies of tiny organisms called coral polyps. Most living coral polyps within a colony are smaller than the eraser at the end of a pencil.³ A tiny coral colony that's just getting started might have only a handful of polyps. An enormous colony will have many thousands of polyps. Coral polyps may be small, but these invertebrates, millions of them together within a reef, build limestone skeletons so large that some can be seen from space.⁴ Their construction projects, coral reefs, become ocean epicenters of biodiversity, creating habitat for as many as a million species,⁵ including tiny fish like bluehead wrasse that dart between branches of coral, damselfish that defend their territory with intimidating snapping sounds, sharks that loiter in sandy patches between corals, bright orange scallops that use long spines to wedge themselves in reef crevasses, turtles that soar placidly over a reef's colorful diversity, shrimp that make constant crackling noises, snails that cling to sea fans, lobsters that duck into holes, and countless microbes that are unseen by divers.

With thousands of coral polyps in a large colony, and hundreds to thousands of colonies in a reef, corals are powerful. Together, they are strong enough to buffer storm waves, protecting coasts, including our cities and towns, and creating habitat for reef life that prefers calm water protected from the waves. Yet reefs are in trouble today, in large part because corals have been weakened by warming seawater, pollution, overfishing, disease, and other problems (all of which we'll explore in the coming chapters). Some problems, like coral disease, are exacerbated by other environmental changes. Disease has been particularly problematic in the Caribbean—scores of corals have been killed by a disease that debuted in 2014 not far from the park bench and port. And extreme heat is too much for many corals to bear worldwide. For decades, climate change has caused widespread coral bleaching events that have weakened the corals; as I write this, corals are bleaching and dying en masse in particularly hot waters.

It isn't change itself that is problematic for these ecosystem-building organisms. Corals naturally live with change. They cope with cycles of tides and seasonal heat. They also live with regular oscillations between El Niño and La Niña—changes in winds and ocean currents that cause tropical climate and sea surface temperatures to swing warmer and cooler every few years. But these natural sources of change are cyclical—water warms and cools each year with the seasons, for example—unlike problems affecting modern reefs, which only increase over time in an unsustainable way. Predictions about how long coral reefs may hold on in the coming decades are becoming increasingly dire.

To get some perspective amid the current spate of problems, this book looks to the past, exploring the history of coral reefs. Fossil coral reefs make it possible to understand how reefs lived before human influence, helping us understand what today's reefs would be like if we weren't changing their environments. They give us a window into an alternate universe, where coral reefs are subject only to the stressors found in nature instead of the myriad problems humans have caused and continue to cause. This is not just a nostalgic desire to escape the current mess. Looking into the past can also help us understand what reefs need in order to have a future.

Research on the remains of long-dead reefs can also help us understand how reef ecosystems function over hundreds to thousands of years, at much longer timescales than those of studies that observe living reefs. By studying both fossil and living reefs, scientists can gain much stronger insight into how contemporary changes in coral compare to reefs across their long history. Understanding the past and present will in turn help us understand coral reefs' future—and maybe even how to save them. The stories of how past reefs persisted and when they didn't can help us understand how to play the long game: take actions now that will help coral reefs have a brighter future in hundreds and thousands of years, even if the near term is rough.

When they aren't in park benches, the remains of ancient corals and other reef life can be found fossilized in rock outcrops, often on tropical islands and coasts. Most of the fossilized reefs we'll explore in this book lived about 125,000 years ago, a time when there likely weren't even a million humans living on Earth, so they lived with little or no human disruption. We'll also look at the not-yet-fossilized skeletons of coral and other reef life that lived decades to several thousand years ago, which hold a record of both human disruptions and natural environmental changes. These, combined with fossil reefs and modern reefs, add up to a timeline of reefs. By tracking the history of reefs from before human influence to the present day, we can learn what has changed. What went wrong? And when did it start to go wrong? The answers, as we'll see, are often complex, without a single smoking gun.

∼

If you are a fan of dinosaurs, trilobites, or other extinct fauna, even the oldest fossils described in this book will seem very young. This book focuses mostly on coral reefs that lived far less than a million years ago. While that's a very long time from a human perspective, it's nothing in the long span of over 4.5 billion years that is Earth's history (known as geologic time or deep time). The same coral species alive today also lived in reefs at this shallow end of deep time (with a couple of exceptions). By comparing contemporary corals with their relatively recent ancestors in the fossil record, we can easily see connections between the past and present.

While this book focuses on young fossil reefs, there are much older fossil reefs in the world. One of the oldest known fossil reefs sits in the middle of Canada and formed nearly three billion years ago from mounds of blue-green algae called stromatolites in a very different world.[6] There are fossil coral reefs in the US Midwest that formed in the Paleozoic era, approximately 460 to 360 million years ago. In the Guadalupe Mountains of Texas, a huge fossil reef, 400 miles long, is filled with fossil sponges, algae, and other creatures that lived about 275 million years ago, toward the end of the Paleozoic.[7] Reefs have existed for a long time, but it's the geologically younger reefs, with the same species that make up reefs today, that can best inform us about what today's reefs need and what their future may hold.

Earth's history is represented graphically as a stack of layers in the geologic timescale. It's like a map of deep time. The oldest layers, the Proterozoic and Archean, are on the bottom of the stack. The Cenozoic era is at the top of the stack with its most recent time layers: the Pleistocene and the Holocene, the latter of which started 12,000 years ago and appears to end at the top of the graphic. The top of the geologic timescale is the time we're living in now.

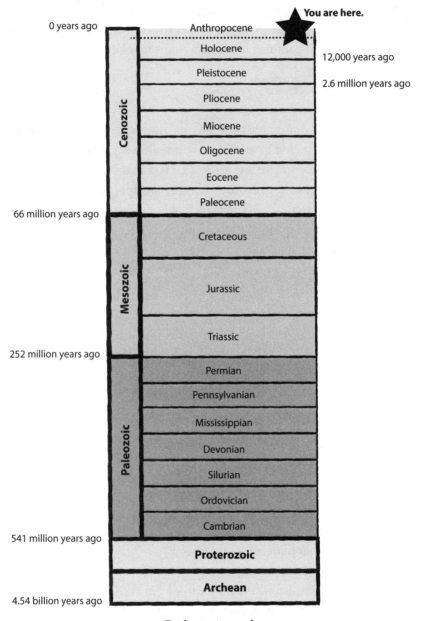

Geologic timescale.

We could be seeing humanity as a part of this long story of the planet, but more often than not, we explore human history and deep time separately. Yet both are made up of seconds, minutes, hours, days, and years. In deep time, the record is usually compressed, making the smaller increments of time indiscernible, but overall, time is time. This makes me want to wander the halls of geology departments and classrooms around the country and the world, adding "you are here" stars to the top of every geologic timescale posted in the style of way-finding maps on hotel room doors.

Our "you are here" star should be in a time period called the Anthropocene, but it's rare to find that word listed at the top of geologic timescales (except for the one in this book). More than two decades after atmospheric scientist Paul J. Crutzen proposed the Anthropocene as a new time period that we've entered because of our enormous effect on the planet,[8] the group of timekeepers that determines the official periods of geologic time, the International Commission on Stratigraphy, decided that, no, the Anthropocene will not become an official epoch of geologic time.[9] This was in 2024 after years of deliberation and research, after a subcommittee voted that we are, in fact, in the Anthropocene,[10] and after a four-year search for an appropriate start date landed on the mid-twentieth century.[11]

Perhaps the Anthropocene will eventually be added to the top of the stack of layers in the official geologic timescale. But even if it isn't, the Anthropocene still describes this time when human impacts are a dominant force on the planet, changing the climate and ecosystems worldwide, and this makes it a useful concept for this book as we plow through the many thousands of years at the current end of geologic time in search of coral reefs. No matter when, specifically, it started, we are most likely still at the beginning of the Anthropocene since it does not seem possible that humans will stop being a dominant force anytime soon, which begs the question: What will the rest of this time layer be like?

This book looks to answer this question for coral reefs, understanding how the story of coral reefs will likely continue through the Anthropocene given what we know about their past in deep time, their more recent history, and the present-day world at the current end of geologic time.

～

I spent my childhood searching for shells on pebbled Cape Cod beaches. I spent my twenties doing much the same as a scientist working on my PhD, although most of those shells were within fossilized coral reefs on tropical islands. Over several years, I headed to the Bahamas for weeks to study the shells and skeletons of marine invertebrates that lived in coral reefs. I was a professional shell collector. And because most of the shells I studied were fossilized, I was also a time traveler, living in the past. I daydreamed about business cards that announced my twin specialties: Dr. Gardiner, Shell Collector and Time Traveler.

In the years since I left research science for the world of climate and geoscience education, I've focused on scary prospects for the future, like climate change and other looming disasters, rather than on the geologic past. For the most part, geologists and paleontologists look back in time, and climate scientists look forward (although there are climate scientists who study ancient climates and geologists who project into the future). I like both the past and the future, perhaps because I tend to have trouble living in the present. There's an anxiety to the present, especially when it's filled with unprecedented environmental problems like the warming climate and its ill effects.

Researching geologically young fossil reefs and seeing their living counterparts in decline just offshore, I found it impossible not to wonder about the ecosystem's future. And it was apparent to me that understanding the past can help us put the present-day trouble into context.

What I've learned while researching and speaking with experts for this book is that the future won't necessarily continue on a predictable trajectory. As we'll see, coral reefs are facing grim times and there will continue to be losses, but there's evidence that we can help turn this around. If we have the power to crumble reefs, we also have the power to fix them, and myriad scientists and conservationists are finding creative ways to apply our understanding of past and present coral reefs to help them have a future. The corals in the Frost Museum's aquarium are part of one such project. This book takes the stance that we can do a lot to improve the prospects for future coral reefs, especially if we use what we know about how reefs have survived in the past and take action on climate change.

In many ways, this book is a search for examples of coral reef perseverance. Stories of how reefs have survived in the past can hold clues to how reefs might survive climate change and other environmental challenges, now and in the future. As we explore these stories from the past, we'll find that over long timescales—millennia or even hundreds of thousands of years—coral reef ecosystems have been able to persist even when individual reefs could not, relocating and rebuilding where corals could survive, lying low when conditions weren't great. Yet this doesn't appear to be the case over short timescales, like the past several decades.

On short timescales, this perseverance is resilience. In ecology, the term "resilience" describes the ability of an ecosystem to recover from disturbances. For example, a coral reef with resilience is more likely to bounce back after a hurricane, even if it takes a few years. Resilience also describes the ability of animals to bounce back—the ability of coral to recover instead of dying after spending weeks or months in overheated water, for example. When it comes to resilience, we have something in common with corals: resilience also describes our ability to recover from stress or trauma and avoid post-traumatic stress disorder (PTSD). At a larger scale, resilience also describes the ability of our cities and towns to recover from extreme weather events and climate impacts.

There are plenty of unknowns about whether corals and reef ecosystems have enough resilience to recover from current challenges and future climate change. We don't know whether they'll be able to persist. Some say coral reefs may be among the first ecosystems to become extinct.[12] Yet the fossil record shows that reefs have been able to survive tough times over the long term. Learning what happened to reefs in the past can help us understand coral reef resilience and persistence, how the world shaped them, how they shaped the world, and what we need to do to help corals thrive now and in the future.

~

In this book, we'll explore fossil reefs at the fringes of tropical islands and dive into modern reefs to connect the past and present. In chapter 1 we'll look at the big picture of how coral reefs and other tropical marine life create limestone that preserves an often-detailed fossil record. This record in rock is what allows us to track reefs through time. Then in chapter 2 we will explore how the idea of uniformitarianism—that the present is the key to the past—helped lay the foundational understanding of coral reefs before Anthropocene reef declines started. As we'll see in chapter 3, the ecological patterns of modern reefs are also found in fossil reefs, an example of uniformitarianism, although living reefs have changed so much that the present is often no longer the key to the past.

We will then step back in time to explore disruption in reefs. This includes the tumult in modern reefs and what dead coral skeletons on the seafloor can tell us about changes in reefs (chapter 4), evidence of Caribbean reef transformation over decades and centuries, preserved in history archives and subfossils (chapter 5), and how reefs coped with environmental change long ago (chapter 6).

Evidence of how reefs have been able to persist over the long term can help us understand how the same strategies could allow reefs that have been decimated to someday return. In chapter 7, we'll explore how one of these strategies, protective refugia where corals shelter in place during difficult times, may help coral species survive. And in chapter 8 we'll see how reef restoration projects are creating human-built refugia, increasing the number of safe places to help reefs persist. In chapter 9, we'll explore other strategies—adaptation and acclimatization of corals to warmer temperatures—which could help corals survive as the climate warms, and how reef restoration projects have been speeding up this process through artificial selection. We'll then consider the possible future of coral reefs based on what we know of their past and present in chapter 10.

Some of the most well-studied fossil reefs in the world are in the Caribbean, which is one reason why this book tends to have a Caribbean focus, although reefs—past and present—from the Indian and Pacific Oceans and the Red Sea make appearances too. Fossil reefs in the Caribbean tend to be well studied because they are often found on land, which makes them easier for scientists to access and study than fossil reefs deep underground or underwater. Also, the first modern reefs to decline are in the Caribbean, with ecological transformations starting decades to centuries ago, which allows us to now look at this decline, and what went wrong, with the benefit of hindsight.

Understanding coral reefs' past, present, and projected future is a multidisciplinary effort. Tracing reef stories from the past to the present gets geologists, paleontologists, paleoecologists, and historical ecologists involved. Marine biologists, ecologists, and conservation experts typically focus on living corals and reefs. Since corals make limestone, geologists do research on living reefs too, not just fossilized ones. Plus, because corals are sensitive to their physical environment, oceanographers, chemists, and climate scientists are also involved in reef research. Many of the scientists highlighted in this book work in interdisciplinary ways to understand reefs. I find hope in the fact that so many experts are looking at questions from many different perspectives to help reefs have a future.

As with most environmental challenges we're currently facing, witnessing both the vulnerability and resilience of coral reefs today provokes feelings of both loss and hope. But knowing what has helped build reef resilience in the past lets us understand what can be done to help corals and other reef species. Just as it takes large numbers of animals to build a reef, it will take large numbers of people to help reefs survive an increasingly inhospitable world.

## CHAPTER 1

# WELCOME TO THE ROCK FACTORY

When alive and underwater, brain coral is often a bulbous mound of yellow ocher, but the brain coral I'm standing on is fossilized, a grayscale skeleton within a rock. It's on land, at the edge of a tropical island with an ocean view. The coral's mazelike surface, although flattened by erosion, makes it stand out from the crowd of other fossil corals in the outcrop—some branching, others rotund—and a medley of smaller fossils like seashells. It's about the size of a beanbag chair but made of rock, so far less cozy, and below it are layers of limestone three to six miles thick.[1] When it was still alive about 125,000 years ago, this brain coral was a small part of a huge limestone rock factory, one of many limestone-generating organisms that created this vast rocky landscape. And now a new generation of marine life, possibly including the descendants of the fossil brain coral, is producing limestone in the shallow water offshore.

A fossil brain coral is an ideal place for a graduate student in a straw sun hat, reeking of sunscreen and bug spray, to take a break from identifying species within a fossil reef for her dissertation research. Because it looks like a brain, I imagine it's a better thinking spot than any other coral skeleton in this fossil reef. And it's also a place to get some perspective because from the fossil brain I can see others, alive, in the shallow water nearby, coping or struggling with the numerous environmental changes that we've caused, problems that this fossil coral never had to

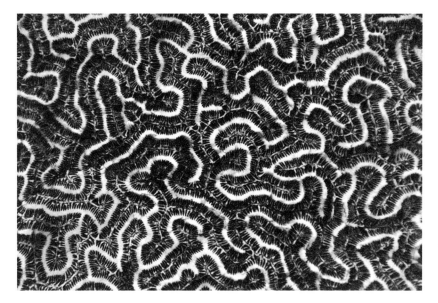

Close-up of fossil brain coral (*Pseudodiploria strigosa*). Photo by James St. John, CC-BY-2.0.

navigate. We will get to those Anthropocene problems eventually in this book, but for now our focus is on the past and what fossil reefs can teach us. In fact, a remarkable amount of information about the ecosystem can be preserved in a fossil reef. That's what I learned while doing research in the Bahamas and is why this story begins there.

I first found my brain coral thinking spot over two decades ago when I was researching fossil and modern reefs as a graduate student on San Salvador, a far-flung Bahamian island. If you find yourself traveling to San Salvador in one of the small charter flights that come in from Fort Lauderdale, you will likely see the island clearly out the front windshield of the plane as you start to descend. If you are in the copilot's seat (told repeatedly, by the pilot to your left, not to touch *anything*), you will have a particularly good view of the low-lying island. But if you are on a commercial flight, looking out a window on the side of the plane, you will see no evidence of land as the plane starts to lose altitude. The water out the window will be the dark blue of the open ocean as the plane descends. But then, the dark blue will shift abruptly to turquoise below the plane, a broad expanse of shallow water so clear that it looks no deeper than a bathtub. Beneath the shimmering surface, it's possible to see the seafloor with its coral reefs built by millions of tiny coral polyps, its rocky rubble and rippled sands. This is where limestone is made.

The plane will continue to descend so low that it will be possible to identify coral species out the window. You'll hear the landing gear deploy long before you see any land to land upon. Just when you become convinced that the plane is making a water landing and wonder how often those inflatable slides at the emergency exits are inspected, the rocky edge of San Salvador and the road that hugs its west coast will appear. The landing gear will make contact with the tarmac. The large leaves of tropical shrubs and coconut palms will wave their greetings.

Many of the rocky areas along San Salvador's west coast, including the one I'm standing on, are remnants of what was once on an ancient seafloor, including coral reefs. The stubby cliffs of drab gray limestone that hold the remains of a coral reef often have choppy scalloped points at their surface that shred the soles of my sneakers. This place doesn't look like the tropical paradises pictured to advertise vacations, but it is a good place to find fossils. Fossils fill the rocks—coral, shells, the burrows of shrimp, and even the remains of tiny encrusting organisms—all of which were once alive on the seafloor, at home in a coral reef.

The fossil reefs here are not an anomaly. There are numerous fossil reefs throughout the tropics. Many of them hug the coasts of tropical islands like this one. Some are on mainland coasts. And there are likely many fossil reefs deep underwater that lived when sea level was lower, although we know less about those since they are hard to reach. Fossil reefs are portals into the past, recording the history of coral reefs through time.

Because of the way limestone forms in the tropics, fossil reefs can hold a detailed record of their history. Most corals cement in place, so they often fossilize in the same spots where they lived. This in situ fossilization becomes the map of an ancient reef neighborhood, telling us which corals lived where the waves crashed, which lived in calmer water, and whether shrimp and other animals were burrowing through the sediments of the reef or trundling across its surface. Studying a fossil reef gives us information about not just the corals themselves, but the rest of the ecosystem as well.

Tropical limestone forms more quickly than most other types of rocks, which means that it can hold very detailed stories of what was happening in a place over time. And because tropical limestone is often loaded with fossils, it's able to preserve a detailed record of past life. Reefs are one part of a colossal limestone rock production system that operates across the tropics worldwide. In a tropical marine environment where fossils make up most or all of the rocks, it can be hard to separate the study of fossils (paleontology) from the study of rocks (geology). Rocks made of fossils, like all rocks, are made of minerals, but unlike the minerals in other rocks, these minerals began as a part of organisms like corals that lived and then died, leaving their skeletons behind. Shells and skeletons, lots of them, make up the bulk of the gray limestone rocks on tropical islands (although many of those shells and skeletons are ground into sand before being cemented into rock). The process continues to this day. A new generation of creatures are living on the tropical seafloor now, creating skeletons that might become parts of new rock. Look at the turquoise water off a fossil coast today, and you'll see where tomorrow's rocks are forming thanks to millions of little animals and other life.

To form the shells and skeletons that become limestone rocks, organisms gather the building blocks they need from seawater. Nearly all take calcium ions and bicarbonate out of seawater to make the calcium carbonate ($CaCO_3$) that forms their skeletons. Calcium carbonate has two common mineral forms—calcite and aragonite—which are made of the same elements but have different crystal structures. Organisms make their shells and skeletons with these minerals, and then these shells and skeletons become cemented together into limestone rock. Because the shells and skeletons in limestone rock are made out of calcite and aragonite minerals, limestone, calcite, and aragonite are all made of the same stuff: calcium carbonate. A small amount of calcium carbonate makes up a clam shell. A large amount solidifies into an outcrop like the one I am standing upon in the Bahamas.

Like other sedimentary rocks, limestone contains numerous grains cemented together. In the limestone that makes up a fossil reef, huge, boulder-sized coral skeletons are cemented in place alongside small sand-sized grains.

~

Corals are able to create impressively large skeletons despite their small size, particularly colonial species (and most corals are colonial, with numerous individuals living together). A single coral animal, called a polyp, is often less than an eighth of an inch (3 mm) across.[2] Some species have polyps as wide as a penny, and a few can be much larger. Coral polyps are simple—mostly stomach with a mouth and tentacles. They have no brain, no eyes, no long-range plan. Yet these delicate invertebrates, many thousands of them in large colonies and hundreds to thousands of coral colonies within reefs, create the ocean's epicenters of biodiversity, possibly living for thousands of years and acting as bulwarks to protect coastlines from the waves of fierce storms. All together, they are powerful.

Those who dive or snorkel in reefs know to never stand on coral, to not even touch its surface because what may look like boulders on the seafloor are actually living animals. On the surface of a living coral colony, the polyps live side by side. Even a light touch can disturb them. They are relatives of sea anemones and, close up, often look like tiny anemones with a mouth at the center of a sac-shaped body that is surrounded by a ring of tentacles that help the coral polyp pluck zooplankton and other food from the water and get the morsels into its mouth. Some, like star coral polyps, are round when seen from above. Others, like brain coral polyps, are elongated, following the mazelike paths we can see in their fossil skeletons. The polyps within a colony are genetically identical clones connected by tissue that bridges between them.

Below the soft bodies of the living coral polyps is the colony's skeleton. Each coral polyp lives at the top of its skeleton as if it's in the penthouse of its own apartment tower. As it grows its skeleton, it adds more stories to the building below (called the corallite) and remains in the

Coral polyps (living star coral)

Skeleton (tops of corallites)

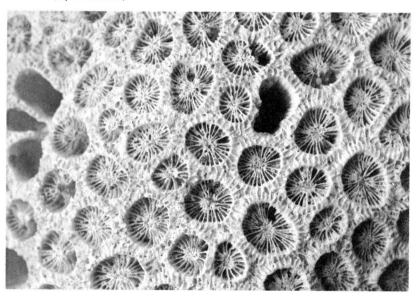

Boulder star coral (*Orbicella* sp.) alive and with polyp tentacles extended (top) and dead, showing the tops of corallites (bottom). Top photo by author, bottom photo by pxhere.

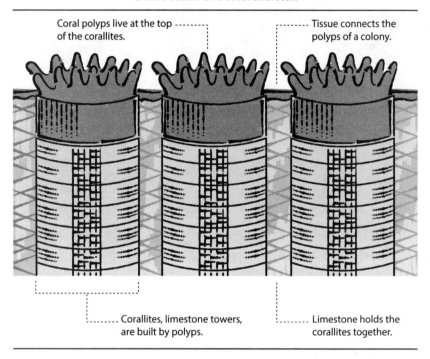

**Cross section of a coral skeleton**

Coral polyps live at the top of the corallites.

Tissue connects the polyps of a colony.

Corallites, limestone towers, are built by polyps.

Limestone holds the corallites together.

penthouse at the top. All its neighbors in the colony do the same in corallites built side by side. The ability to grow their skeletons upward can help corals remain within shallow, sun-drenched water when sea level rises. The brain coral that became my thinking spot during field research used to grow in this way, but its penthouses had been vacant for about 120,000 years. Dead coral colonies no longer have polyps, so all that's left behind is the projects they worked on during their entire adult lives: their elaborate skeletons, which may become fossils.

While coral colonies grow their skeletons, they also create new clones, adding new apartment towers topped by new coral polyps that are genetically identical to their neighbors. A coral colony can grow like this for hundreds, even thousands of years, or until the environment changes and the coral is no longer able to survive. If a branch of a coral colony breaks off, it may become a new outpost, cementing to the seafloor and adding polyps to grow another colony out of the fragment.

That's one way that corals reproduce asexually. Corals also reproduce sexually, most through spawning: sending gametes into the water that will be fertilized when they run into gametes from a different coral of the same species (corals can't sexually reproduce with their clones).

Most corals need to live at shallower sunlit depths because within the polyps are numerous single-celled algae that need sunlight. The algae, called zooxanthellae (usually species in the genus *Symbiodinium*), are key members of the coral's enormous microbiome, the community of microbes within the coral that includes bacteria, archaea, protists, fungi, and even viruses. The zooxanthellae get a safe space to live inside the coral and the nutrients they need; the coral gets food (sugar) produced by the zooxanthellae through photosynthesis, which supplements the zooplankton they hunt with their tentacles. This relationship is symbiotic—beneficial for both the corals and zooxanthellae.

It's because of zooxanthellae that corals are able to create so much limestone: that extra source of nutrition helps coral polyps build more floors onto their apartment towers, growing the size of the colony and the structure of the reef. Unfortunately, when water gets too hot, the zooxanthellae may create toxins, causing corals to expel them. This is coral bleaching. Without their algal symbionts photosynthesizing, corals often don't have enough food. They grow weak. Some corals don't survive bleaching, but some do recover, taking in new zooxanthellae and regaining their color when waters cool. Despite the potential for recovery, corals grow their skeletons more slowly when they are stressed and bleached. For a reef to be geologically growing, more limestone needs to be added than erodes away. When bleaching throughout a reef slows coral growth, which is likely if it happens repeatedly, the reef, while still alive, may no longer expand its rocky framework quickly enough to outpace erosion. The structure can weaken and be unable to grow upward fast enough as sea level rises, which over many years can cause corals to die.

Looking closely at the gray limestone that makes up a fossil reef like the one on San Salvador is like scuba diving on land except that, well, you stay dry, and all the marine life died millennia ago. You'll find dome-shaped coral fossils—some the size of a basketball, and others the size of a Volkswagen. You'll find branching corals in the rock too, including large amounts of elkhorn and staghorn coral, so named because they resemble the antlers of hoofed mammals (although elkhorn coral looks more like the antlers of what North Americans call a moose, rather than an elk). Of the coral species found in Caribbean fossil reefs from the late Pleistocene, all but two are still alive today, which can make a fossil reef look like a skeletonized version of a modern reef.

At first glance, the corals are the most prominent feature of the fossil reef, but reef limestone is more than just corals. Get down on your hands and knees and take a closer look, and you'll see that there are a lot of other fossils, much smaller, between the corals. Corals often do not fit together like falling Tetris® blocks. Their branching and rounded shapes leave space between. It's in those spaces that you'll find evidence of coral fragments, clam and snail shells, urchin spines, and even tiny fish teeth. But not all reefs have spaces between the corals. Such nooks and crannies may be more common in Caribbean reefs. Caribbean reefs have been described as garbage piles because they have ample spaces for reef detritus to accumulate.[3]

The largest fossils you'll find between corals are the shells of clams and snails that once lived on, or just below, the seafloor. Clams and snails build calcium carbonate shells and usually also add a small amount of protein to their shells for reinforcement, like rebar in concrete. Fossilized, their shells are white or gray, drained of their once vibrant colors. In the Caribbean, all the clam and snail species we find as late Pleistocene fossils still exist today in the shallow water offshore.

Look even more closely at the limestone, in between the clam and snail shells, and you'll often find sand grains that look like flakes of uncooked oatmeal. Some flakes are the size and shape of "old-fashioned" oatmeal and others, eroded at their edges, resemble instant oatmeal. These flakes are from a type of algae called *Halimeda*, which produces calcium carbonate within its tissues.

*Halimeda* is one of the most prolific sand producers on the tropical seafloor in and around coral reefs worldwide.[4] Alive, *Halimeda* often looks like a little green tree with branches made of strings of flattened emerald beads. Snorkel in shallow water and you'll find forests of these trees on the sandy seafloor. The sand at the base of an algae forest is mostly the flake-shaped remains of dead algae, not unlike leaf litter on a forest floor. You'll also find evidence of other types of algae that produce calcium carbonate on the seafloor today, too. Some look like green lollipops. Others look like ginkgo leaves. And then there are coralline algae, which encrust over surfaces, gluing parts of a reef together with calcium carbonate.

There are also very small fossils, such as the remains of single-celled protists called foraminifera, which create little shells, often the size of sand grains. Look at forams (as they are called) under a microscope and you'll find that each has a calcium carbonate skeleton, often with a combination of elaborate and delicate bubbles, spirals, and points. The shapes are reminiscent of snail shells, although they are not related. Quite impressive for unicellular creatures. Preserved in rock, they become some of the world's tiniest fossils.

Nearly all limestone in tropical oceans is made by living organisms. In some places, limestone appears to be forming inorganically; however, recent research suggests that in many of these cases microbes are actually pulling the strings, creating an environment where calcium carbonate is likely to form from seawater.[5] Such is the case for unusually round sand grains called ooids. At first glance, ooids may look like any other sand, but they roll as easily as tiny ball bearings. Look at them with a magnifying glass and each grain will look nearly like a pearl. When split in half, an ooid shows concentric layers of calcium carbonate, like tiny jawbreaker candies. And ancient ooids may be evidence of ancient microbes.[6]

On San Salvador, there are limestone rocks made of millions of ooids that are the same age as the fossil reefs, indicating that the environment at that time may have looked much like the blanket of ooids today on the shallow seafloor north of Andros Island, Bahamas, around the Joulter Cays,[7] little islands inhabited by numerous birds and no humans. Rolling on the seafloor, the ooids mound up into ridges in some areas and are bisected by channels in other areas. Looking at the myriad ooid-filled Pleistocene rocks on San Salvador, we could imagine a similar scene and perhaps even similar microbes.

While healthy coral reefs are usually home to numerous fish, you are unlikely to find fish bones in the space between corals of a fossil reef. They are typically less durable than coral skeletons and seashells and break as they are tossed by waves.[8] Fish teeth are often preserved, but fish add the most bulk to limestone with their poop. Along with other excretions, fish poop contains calcium carbonate, which creates lime mud on the seafloor.

Tropical marine fish are prolific poopers. Scientists estimate that fish in the shallow waters of the Bahamas produce over 6,700 US tons (over 6 million kg) of lime mud per year,[9] which adds up to 14 percent of all the lime mud in the Bahamas. In some shallow marine habitats, fish poop makes up more than 70 percent of the mud.[10] The areas on the protected side of coral reefs are particularly poop filled. These are often places with beautiful, calm water, round heads of coral, sea fans, and darting fish. They are some of my favorite places, but it's hard not to think about all that poop now that I know about it.

Why do fish have lime mud in their poop? They need to stay hydrated but have access only to seawater. Fish are able to ingest seawater and absorb only the water, leaving the salt, calcium, carbonate, and other elements behind. Tiny mineral crystals form in their guts, which are excreted along with the other, more crappy waste.

Parrotfish have another way of making sediment. Parrotfish are so named because their jaw resembles a beak and their festive tropical colors are not unlike those of their namesakes. Found in reefs worldwide, they are an important part of coral reef ecosystems, as we'll see in later chapters, because they eat mostly macroalgae, or seaweed—types of algae that don't make calcium carbonate like *Halimeda* and coralline algae. When a parrotfish swims toward an algae-covered rock in a reef with mouth agape, its goal is to eat the algae. But most parrotfish species aren't that detail oriented and wind up consuming fragments of rock along with their food. Parrotfish poop contains the fragments of rock they eat. This can really add up. According to *Life Sculpted* by Anthony Martin, which devotes a chapter to parrotfish poop, there are parrotfish that can eat and poop up to 2,200 pounds (1,000 kg) of sediment a year.[11] The sediment grains are larger than the grains of mud excreted by other fish and smaller than typical grains of sand.

Parrotfish can crunch rock because they have a second set of jaws deep in their throats outfitted with numerous teeth for grinding and crushing. These somewhat tubby fishes, each with a slight Mona Lisa smile, zoom around a reef, rowing their pectoral (side) fins like oars, and periodically chomping at the limestone structure of the reef.

My experience scuba diving has taught me that parrotfish are some of the noisiest residents on a coral reef. The crunch of rock can be louder than the crunch of Doritos®, and they are eating constantly, unless they are asleep (which, incidentally, is adorable). Noise isn't fossilized, but there is evidence of that noise preserved in limestone rocks that include parrotfish teeth and sediment from their poop.

~

Over time, the limestone created by multitudes of coral, mollusks, fish, algae, forams, and other life really adds up. Step back to see the broad view and you'll find limestone added layer after layer, millennium after millennium, eventually creating colossal underwater platforms, each shaped somewhat like a mesa with a flat top and steep sides. The shape of these platforms affects where reefs form. Corals with symbiotic algae need to be within reach of sunshine, so they are found on the shallower parts of a platform. And the reverse is true as well: reefs affect the shape of the platform, at least its shallowest parts.

Many tropical islands are on the top of these limestone platforms, and, looking out from one of those islands, you can see evidence of the mesa-shaped platform below the sea surface. If you were to look off the coast of San Salvador, you would see a view over the ocean and atmosphere with three horizontal stripes, each a different shade of blue, a composition not unlike one of Mark Rothko's color field paintings.

The top stripe, farthest away, is the light blue of the sky, which is especially pale when humidity is high. The middle blue stripe, which is about four to six city blocks offshore from San Salvador's west coast, is dark and uniform. This is where ocean depth increases sharply from about 40 feet to over 6,000 feet at an underwater cliff. Not every limestone platform has a near-vertical drop-off like San Salvador, but the sides of a platform are usually steep. In general, steep drop-offs are rarely found beyond the tropics, and this has interesting implications for where corals could survive at times when sea level was low, as we'll explore in a few chapters.

Snorkeling over the edge of this cliff gives some people vertigo, but it makes me feel like I'm flying. The seafloor drops and I remain high above, breathing a bit like Darth Vader through a snorkel and looking down through my mask into the abyss. The water may look crystal clear, but it's not possible to see the seafloor past the drop-off. It's too far down. There are thousands of feet of water below, and light, scattered and absorbed, fades with depth. Particles in the water also obscure the deep like a fog. Staring into the void always makes me wonder how many animals could be directly below me. Could there be a whale shark down there? A giant squid? It's possible. And yet I see none of these creatures in the water below, just endless dark blue.

The lowest blue stripe of this tropical Rothko, the turquoise one, is closest to shore. This bright turquoise stripe is the shallow top, or shelf, of the massive underwater platform. Along with the shallow parts of the steep slope, this is where most animals, algae, and microbes are creating their shells and skeletons, some of which will eventually become solid limestone. In these areas, the water is clear enough for sunlight to get to the seafloor, allowing zooxanthellae—the algae within corals—and coralline algae to photosynthesize. This shallow platform top is the heart of a rock factory, where most limestone rock production happens, and it can hold detailed records of coral reefs through time.

While it might seem counterintuitive, limestone platforms often grow taller by sinking, which happens as the Earth's crust drops a bit lower over time—either because the underlying plate drops lower as it warps, stretches, or cools, or because the layers of limestone within the platform compact with the weight of layers above. Some platforms sink rapidly and some sink very gradually. If corals and other organisms grow upward at a rate similar to the rate of sinking, their skeletons and shells fill the space, so that the water of the platform top (the shelf) remains

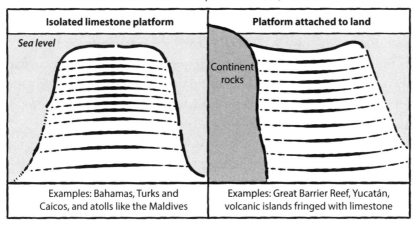

**Two types of limestone platforms**
Layers of limestone form mainly at the top over time.
Platforms can have areas of the top above sea level, which are islands.

at around the same shallow depth while the distance between the top and bottom of the platform increases, sometimes for millions of years. Global sea level rise contributes too because it creates more space for corals and other life to fill. If a limestone platform were a cake, it would look like layers were being added on top as the cake plate dropped bit by bit. On San Salvador, Bahamas, this process has produced an undersea limestone cake several miles thick that is made from the skeletons of marine life.

There are two types of limestone platforms: those that are attached to continents or islands made of other types of rock, and those that are isolated on their own.[12]

Volcanic islands in the tropics, such as in the southern Caribbean, Hawaii, and the Pacific, often have attached limestone platforms. This brings rock made from lava close to limestone made from shells and skeletons and can create beaches of black lava rock mixed with white limestone pebbles. But eruptions can also shut down limestone production temporarily, when lava, ash, or sediment covers seafloor marine life.[13]

Coral reefs, the major infrastructure on limestone platforms, come in three types—barrier reefs, fringing reefs, and atolls. The world's largest barrier reefs are on limestone platforms connected to continents.[14] These include the limestone that underlies Australia's Great Barrier Reef and the Mesoamerican Barrier Reef off the east coast of Central America. In both cases, you'll find the bulk of the reef parallel to the coast but at a distance away. Smaller barrier reefs encircle islands.

Fringing reefs are close to shore (fringing the coast), and you can find them on limestone platforms attached to land as long as there isn't too much sediment eroding from the land, which can smother a reef. You'll also find fringing reefs around islands on isolated platforms.

You will find coral atolls only on isolated platforms. Most are in the Pacific and Indian Oceans and have a roughly circular shape that follows the outer edge of the shallow shelf. Charles Darwin discovered that their curious circular shape forms as islands sink, as we'll see in chapter 2.

Living reefs are often at or near the top of limestone platforms, but fossilized reefs could be anywhere: in the deep ocean on the steep slope of a platform, deep underground in ancient layers of limestone, or even on land like the fossil reef on San Salvador. The likelihood of stumbling across fossil reefs on land at the top of a limestone platform depends on how the underlying plate has moved over time and also how sea level has changed through the most recent ice age, which started about 2.6 million years ago and is likely still ongoing.

Over geologic history, Earth has gone through periods of warmth called greenhouse periods, and relatively cool periods called ice ages. In general, ice ages are times when global average temperature is cool, but it's not cold the whole time, or at least it hasn't been during the most recent ice age. Temperature has swung back and forth between colder and warmer during the ice age. Colder ice age times are called glacial periods and warmer times are called interglacial periods. Note that popular culture has dubbed the last glacial period, which occurred about 20,000 years ago, as the ice age. Because this book explores reefs from various glacial and interglacial times, we'll use the term "ice age" to refer to the longer timeframe of the past 2.6 million years. We have been in an interglacial period since the start of the Holocene, nearly 12,000 years ago, but human-caused climate change has likely disrupted these cycles.[15]

While the water in the tropics has stayed warm enough for corals to survive during this ice age, oscillating sea level has disrupted reefs. During glacial periods, global cooling causes ice sheets to grow near the poles, trapping water in ice that would otherwise fill the ocean, which leads to lower sea levels. At the height of the last glacial period, about 20,000 years ago, ice sheets were so large that they extended from the Arctic as far south as New York, Ohio, and Illinois. These ice sheets were so thick that the top of the ice was higher than New York City skyscrapers.[16] That's a lot of water trapped on land instead of in the ocean, which is where much of it is now. During interglacial times like today, the sea level is high, both because less water is in glaciers and ice sheets and because ocean water expands when it warms.

During the last interglacial period (between 130,000 and 115,000 years ago), sea level was about 20 feet (6 m) higher than it is today,[17] so some low-lying tropical islands that exist today were partially or entirely underwater, and reefs formed in places that are now land. We can find fossilized reefs from this timeframe above sea level at the top of limestone platforms that have been stable—not sinking or uplifting much.[18] This is why the large fossil reef on San Salvador and my brain coral thinking spot are, today, on land.[19] We know a lot more about fossil reefs from this timeframe because they are easier to access than many other fossil reefs.

Elsewhere, shifting tectonic plates have caused land to move up or down as global sea level rose and fell during ice age cycles. For example, in places where islands are uplifting as the tectonic plates underlying the islands slowly rise, you may find multiple fossil reefs on land because the shallow seafloor keeps rising above sea level like an upward-moving conveyor belt. For example, on the island of Barbados in the southeast Caribbean, which we'll visit in chapter 3, reefs and the remains of other seafloor life moved upward onto land by this conveyor belt process over the past million years, and now these fossils cover the whole island.[20]

~

The base of a limestone platform may be tens of millions of years old, or even 100 million years old, but the tops of platforms formed very recently and are still forming where they are underwater right now, which is a reminder that geologic time is not finished. Earth is a work in progress.

If you stand on a fossil reef on San Salvador atop a limestone platform, it's possible to see this work in progress—ecological stories from the past are underfoot while the living reefs, often struggling, are just offshore, dark patches within the turquoise shallows. This place is at the juncture between the past and future during a time of rapid change within the vast history of the Earth. Comparing fossil reefs with the floundering modern reefs, I find it impossible not to think about what reefs of the future may be like—both in the near term, mere decades from now, and in a more distant future, a leap in time that's perhaps as far into the future as the fossils underfoot are into the past.

I made several trips to San Salvador to study the fossil past and precarious present-day life within coral reefs, focusing mostly on clams and snails in reefs, but the corals were impossible to ignore since they form the essential structure of reefs, modern and fossil. To get to San Salvador, I would fly east from Miami or Fort Lauderdale, sometimes stopping in Nassau, the capital city of the Commonwealth of The Bahamas, if I couldn't get a direct flight. Out the window, the ocean below would turn abruptly from navy to the color of a swimming pool as we flew over the shallow tops of limestone platforms. My brain would reorganize, preparing to live in the past with my nose in the fossil record and my mind busy comparing it with the present world.

Even at the altitude of a commercial flight, the waters over platform tops are so clear that underwater sand dunes and patches of seafloor life are visible. What's impossible to see are the millions of animals alive in the reefs, the fish and their limestone poop, the forests of *Halimeda* algae, the millions of ooids, and all the other animals that are collectively responsible for the limestone produced in the shallow turquoise water.

Limestone platforms underlie the islands of the Bahamas. The largest—at about 300 miles across—is called the Great Bahama Bank, which seems like the name of a financial institution. From overhead, its platform top of turquoise-blue water has a globular, irregular shape with its western edge about 50 miles east of Miami, Florida. The state of Florida is perched on a limestone platform too.[21]

Farther afield, a platform is attached to the east coast of Central America, and smaller platforms fringe other Caribbean islands and coasts. There are numerous limestone platforms in the seas north and east of Australia and south of Southeast Asia where the Indian and Pacific Oceans meet. These include Australia's Great Barrier Reef and the extensive reefs around Indonesia, Malaysia, and Papua New Guinea. Limestone platforms also make up the Maldives and other remote islands at tropical latitudes of the Indian and Pacific Oceans. There's also a limestone platform within the Red Sea.

A flight from Florida over the Bahamas crosses above some of the most prolific limestone rock factories. Limestone production in the shallow ocean is fastest at tropical latitudes because seawater is warm enough for corals to thrive (yet, we hope, doesn't get too hot) and is oversaturated with calcium carbonate. Limestone forms elsewhere in the ocean, but not nearly as fast as in the tropics.

While much of the Bahamas is on vast limestone platforms, some of the islands are on small, isolated limestone platforms. San Salvador is on a small platform some distance east of most other Bahamian islands, so the water below the plane returns to a dark blue monotony for a while before its bright platform appears.

San Salvador and its platform may be small, but a lot of research happens there. The charter flights I took to get to San Salvador were filled with scientists—biologists, geologists, paleontologists, archaeologists, or others who explore the Earth and life on it. Some investigated the iguana population while others researched Lucayan archaeological sites. I was there for the coral reefs and seashells—fossil and modern. On commercial flights from Miami, I would see a mix of locals returning home, tourists heading to the large Club Med or smaller encampments, and just a handful of scientists. By the time we would arrive in San Salvador, most of the scientists would have undergone a butterfly-like transformation from university professors and students concerned with grading exams and planning next semester to the fieldwork version of themselves. For paleoecologist Sally Walker, my dissertation adviser at the University of Georgia, this transformation included a sun hat atop her head so broad that she had shade to spare. After stepping out of the plane, donning her sun hat, and heading down the narrow staircase to the tarmac, she would extend her arms as wide as her hat, a particular grin spreading across her face that I have seen only in the tropics. For other scientists, this transformation might include bandannas around necks or a khaki vest with lots of pockets and a backpack. Large water bottles, bug spray, and sunscreen were usually standard accessories.

We would walk across the runway from the plane to a cheery yellow one-story building—the terminal of the San Salvador International Airport. It is the only place to go. A large Bahamian flag, turquoise and yellow with a black triangle along one side, waves above a large, inviting porch facing the runway. Through the porch and a set of French doors, the passengers from the flight would line up and wait for Immigration and Customs.

I was always perplexed about how to fill out the disembarkation/embarkation card required for arriving passengers. The card requests that visitors specify the type of accommodation where they'll be staying during their visit to the Bahamas. None of the options listed (hotel, rental villa, one's own property, a timeshare, a friend's or relative's home, or a private boat) really fit with the repurposed midcentury US Navy station where I would be staying. I was glad the form included an "other" box, which seemed like the best choice.

Another question on the card asks visitors to specify the purpose of their trip. Options include a wedding, a honeymoon, a vacation, visiting casinos, doing business, or "other." The purpose of my visit was to decipher the stories of young rocks and fossils to understand where tropical reefs have been and where they are going. I was visiting San Salvador to learn from past and present-day marine life.

Again, I was grateful that the Bahamas Department of Immigration included the "other" checkbox.

According to US Immigration and Customs Enforcement, there are only two reasons for travel. "Business or pleasure?" a US immigration agent would ask each time I returned to Miami.

"Ummm . . . it was pleasurable, but it wasn't a vacation," I would say wistfully. Compared to the immigration checkpoint—a lackluster windowless hall full of circuitous rope-lined pathways—the sunny outcrops of fossils and the tropical ocean I had just left seemed like another world. I learned that answering "business" instead of "pleasure" would cause a deluge of follow-up questions about the commodities I was buying and selling. Field research is not widely acknowledged as a reason for traveling, and it is neither a vacation nor a business trip, at least not in the capitalist, commerce sense of the word.

In San Salvador, I would hand my card, "other" boxes checked, to the agent at the immigration desk with my passport and await questions about the nature of my trip. But more often than not, the agents working immigration at the San Salvador International Airport had no questions. They would look at my "other" answers and smile knowingly. I wasn't the only "other" in line. They would wave me through to customs. Beyond, out the far door of the terminal, an old blue truck was waiting to take the scientists who had checked the "other" boxes and our mountain of luggage on a bumpy ride to the field research station. After a day spent traveling across geography, I would be ready for time travel, exploring life and environments of the past preserved in limestone and present-day environments where life is still making limestone atop this isolated platform.

# CHAPTER 2

# THE PRESENT AS THE KEY TO THE PAST

The utilitarian cement block buildings of San Salvador's Gerace Research Centre were built in the early 1950s for the US Navy, which, with the Air Force and Coast Guard, were on the island during the Cold War with the permission of the then ruling British government. Today, students stay in the former barracks. Faculty and researchers stay in the former officers' quarters—squat buildings that look like one-story motels. Between the blocky buildings are clotheslines filled with swimsuits, towels, and dive skins. Scuba gear sits drying on rocks. The center's largest building serves as lab space, lecture hall, and when needed, the local hurricane shelter. The heart of the complex is the cafeteria, particularly at dinnertime when stories of the day's research adventures flow through the space along with countless mosquitoes and the smell of baking bread.

It's because of the research center that this tiny Bahamian island has become a mecca for natural history research—from biology to geology to archaeology. College courses use it as a home base for learning about the science of the island and its marine outskirts. I first arrived on San Salvador as an undergraduate student on a Smith College field course, which was my introduction to coral reefs and other life on the top of limestone platforms. I didn't know it would be the first of many trips to the island.

Visiting geology and paleontology researchers and students often spend considerable time on San Salvador pondering how its present-day life and environments can help us interpret evidence from the past that's preserved in limestone. Looking at the modern world for clues to interpreting rocks and fossils applies a principle of geology called *uniformitarianism*. Without a time machine, it's impossible to see the processes that formed rocks and fossils, so we look at analogous processes happening today to understand what likely happened in the past.

Uniformitarianism is used to help interpret many types of rock, but it's particularly helpful atop geologically young limestone platforms like the one San Salvador sits on, because modern environments on a platform are often like an answer key that explains the rocks. If you find a fossil or rock in such a place, stand up and look around. You might spot a modern analogue nearby.

There are examples of uniformitarianism all over San Salvador. It's thanks to uniformitarianism that we can look at a fossil reef and understand how the fossils within it once lived based on what we know about their descendants, alive today. For example, my brain coral thinking spot is between numerous corals that, today, prefer to live in shallow water, often with waves crashing overhead, so my thinking spot may have been in an environment like that about 120,000 years ago when the now-fossil reef formed. We can assume that the oatmeal-like flakes of *Halimeda* algae within the fossil reef mean that there were little green trees of the algae on the seafloor in the past, much as there are today. And if we find parrotfish teeth, we can assume those parrotfish were eating algae and some limestone rock, just like parrotfish do today.

We can also learn about trace fossils like burrows and footprints by comparing them to what we find in the present day. Researchers think that fossilized burrows near the fossil reef, which we visited in the previous chapter, were made by a species of ghost shrimp because they are strikingly similar to modern ghost shrimp burrows. Living ghost shrimp can be found on San Salvador near Pigeon Creek on the southeast corner of the island. Their burrows pepper the intertidal zone, making the sand flat look like it's covered with miniature volcanoes, each at a burrow entrance.

When paleontologist Sally Walker, geologist Steve Holland, and I found tiny fossilized footprints in a now-petrified sand dune, we

| Present | Past |
|---|---|
| A living hermit crab marches across modern sand on San Salvador, Bahamas | Holocene tracks on San Salvador made by a hermit crab several millennia ago |

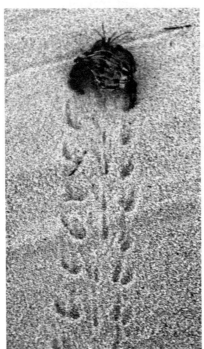 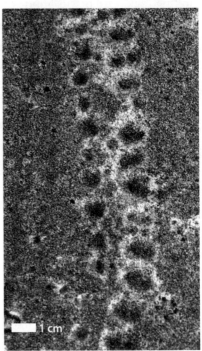

Identifying the fossil tracks we found on San Salvador relied on uniformitarianism, using the present as the key to the past. Photos by Sally E. Walker.

deduced that they were made by a land hermit crab because hermit crabs alive today on the island amble across sand making very similar tracks. They wander the beaches on San Salvador to get water and scout for new shells. Each hermit crab leaves in its wake a line of tracks in the sand with two rows of prints side by side. To determine whether the ancient set of tracks was made by a hermit crab, we applied uniformitarianism, comparing the tracks of living hermit crabs to the fossil tracks.[1]

You can also apply uniformitarianism to figure out how different sedimentary rocks form. In fact, as we'll see shortly, this application is how the idea was first conceived. For example, the rocky coastal cliff at North Point, a short walk from the Gerace Research Centre, is made up of ancient sand dunes. They have the same thin laminations, oriented at an angle, as modern sand dunes. The main difference is that they are now solid rock instead of loose sand. These are young rocks that formed just a few thousand years ago in the early Holocene when sea level was low.

Because limestone rocks can form quickly in the tropics, as we saw in chapter 1, it's almost possible to witness rock making in action on San Salvador. For example, on some beaches, you may find sand cemented together to form rock called beachrock. The beachrock often has a gentle seaward slope, like the rest of the beach, and sometimes a scattering of seashells entombed within it. It might be only a few hundred years old. If you are looking for older beachrock, you can find some that's over 100,000 years old not far from the fossilized ghost shrimp burrows in the fossil reef.

Uniformitarianism is very useful on San Salvador and in other tropical limestone-producing environments where past and present are often side by side, but the principle is also used by scientists interpreting rocks in many other geologic settings. The present day does not always perfectly account for the past—some past events are unlike anything that's happened since—but overall, uniformitarianism has proved to be a useful brick in the foundation of modern geology.

But when it comes to coral reefs as ecosystems today, uniformitarianism has its limits. Recent environmental changes have in many cases caused present-day reefs to no longer resemble the past. For coral reefs in the Anthropocene, it might be more accurate to say that the present *used to be* the key to the past.

As we'll see in upcoming chapters, in some cases uniformitarianism still holds true for coral reefs. And in some cases it doesn't. But overall, this book explores what I like to think of as the inverse of uniformitarianism: the past can be a key to the present (and, we hope, the future). Coral reefs from the past can help us put struggling present-day reefs into context and figure out how we can help them.

Historically, before the Anthropocene turmoil of the last 70 or so years, and around the same time that the burning of fossil fuels was just starting to cause an uptick in carbon dioxide emissions, uniformitarianism was remarkably helpful as scientists sought to understand fossil reefs and how reefs are built. In this chapter we'll dive into human history to explore how, nearly 200 years ago, Charles Darwin used the concept of uniformitarianism to understand how reefs form and grow. Then, in the following chapter, we'll look at how, decades ago, scientists used uniformitarianism to decipher the ancient ecology of Pleistocene fossil reefs, including those on San Salvador.

The idea of using the present as the key to the past can be traced to geologist James Hutton (1726–1797). Hutton lived in Scotland,[2] far from the tropics and the fossil and modern coral reefs that are the focus of this book.

It might seem like common sense to look around at the natural world today and see processes that must also have happened in the past—the way a river transports sediment or the way fossils form out of buried skeletons, for example. But the idea of uniformitarianism hasn't always been accepted. When James Hutton was formulating his idea in the eighteenth century, many Christians in Scotland and elsewhere thought that Earth was about 6,000 years old (according to an Irish bishop named James Ussher, our world was created on October 23, 4004 BCE). If Earth was this young, features like layers of sedimentary rock and mountains would either have to have formed catastrophically fast to fit within that timeline or to have come preinstalled when Earth was created. But scientists, including Hutton, were starting to challenge this idea.

James Hutton didn't start out as a geologist. In the eighteenth century, no one did. It wasn't yet a profession. He went to medical school but then decided he would rather be a farmer than a doctor. Hutton had inherited his family's farmland in the Borders region of Scotland, and although he had no experience with agriculture, he jumped right in, spending two years learning from farmers in England before moving into his Scottish farmhouse and working the land for over a decade. In his spare time, he explored rock outcrops.[3] Like many farmers, Hutton was invested in keeping his farm's soil in the fields, but erosion, he found, was an ongoing process.[4] In the Borders, soil and sediment erode from the land and travel in small streams, eventually ending up at the ocean, where they are deposited on the seafloor. At some point—it's not known when—he speculated that the erosion and deposition he saw happening at the farm might explain the layers of sedimentary rock he saw on his geologic expeditions around Scotland. He was looking at the present as the key to the past.

In 1785, he presented his ideas to the Royal Society of Edinburgh,[5] explaining that rocks are formed and destroyed by processes that remain at work in the world today and that these processes must have been happening for a very long time in order to produce rocks.[6] There was some resistance to the theory, especially since it contradicted biblical interpretations of a 6,000-year-old Earth and also because there wasn't much evidence to back it up at the time.[7]

Three years after he presented his theory, James Hutton found evidence to support it—including a rocky outcrop at Siccar Point, about 40 miles east of Edinburgh. Siccar Point, a headland jutting out into the North Sea on the east coast of Scotland, was not far from his farm. Today Siccar Point can be reached from the James Hutton Trail,[8] which takes one through grassy fields sometimes dotted with cows or sheep, past the ruins of St. Helen's Chapel and old stone walls constructed of the same red and gray rocks that are found in the Siccar Point outcrop.

In the spring of 1788, James Hutton, Sir James Hall, John Playfair, and several farmhands sailed from Dunglass in search of rock outcrops along the coast.[9] James Hutton knew that there were two types of rocks in the area—gray and red—and he wanted to know how they formed. To look at a story of change over time in rocks, he needed to see a cross section that showed the older rocks below and younger rocks above. They found one in the cliffs as they sailed along the coast south of Dunglass. According to John Playfair's account of this trip, the weather was mild and the sea was calm, so they were able to sail close to shore without waves smashing their boat against the rocks.[10] James Hutton was known to be animated even when nothing much was happening, but his excitement could not be contained when he saw the layers of rock at Siccar Point. Eventually, he calmed down enough to explain to the others what they were looking at.[11]

Here is the modern explanation of what's in the Siccar Point rocks. The outcrop's base is made of numerous thin gray sedimentary rock layers that have been tilted vertically and then eroded. Those tilted layers formed about 435 million years ago at the bottom of a deep ocean. Layers of mud were deposited when ocean currents were calm, forming a rock called mudstone. At other times, sand avalanched down the slope and a layer with some sand in it was deposited, forming a rock called graywacke. About 65 million years later, after the gray rocks had been uplifted and tilted by the movement of tectonic plates and then eroded at the top, more sediment was deposited above the tilted layers, including a layer of conglomerate—a sedimentary rock with a mix of large cobbles and sand—and above that, a red sandstone formed in rivers and dunes, its color derived from iron oxide.[12]

James Hutton wouldn't have known any of the dates or specific details about Siccar Point. Plate tectonics was not yet a theory and there was no way to date the ages of rock. But he did realize how the layers of sedimentary rock might have formed through processes that still happen.[13] For example, ripples preserved in the gray rocks were made by moving water just like modern ripples form in a stream or at a coast. If those processes accounted for the Siccar Point rocks, Hutton thought, then the rocks must have formed over vast amounts of time.

Minister and mathematician John Playfair was skeptical of his friend's ideas, but the trip to Siccar Point changed his mind.[14] He became enamored with the concept of the present as the key to the past. He wrote in his account of the trip, "We felt ourselves necessarily carried back to the time when the schistus on which we stood was yet at the bottom of the sea and when the sandstone before us was only beginning to be deposited, in the shape of sand or mud."[15] (Note that old English dictionaries define the word "schistus" as a rock with thin layers. Today a similar word, "schist," refers to a type of layered metamorphic rock.)

At the place in the rock outcrop where the lower gray layers meet the upper red layers, James Hutton could see the vastness of geologic time. That lumpy contact, which is called an unconformity, is where a chapter of the rock record is missing from the stack of rocks. (We'll get to know unconformities within fossil reefs in chapter 6). At an unconformity, either no sediment was deposited for a long time or rock eroded. At the

Siccar Point unconformity, rocks eroded as the top surface of the uplifted and tilted layers gradually wore away, with sediment breaking off and carried away year after year. An unconformity has no thickness, no mass; it is essentially a plane. And Hutton, having observed erosion as a gradual and ongoing process, recognized that this simple plane was a clue to how much geologic time had passed between when the gray and red rocks formed—enough time to tilt the layers vertically and then erode them until the surface was nearly flat. "The mind seemed to grow giddy by looking so far into the abyss of time," Playfair wrote after the field trip.[16]

This trip gave Hutton the evidence he needed to support his theory of uniformitarianism. Although the idea originated in chilly, high-latitude Scotland, it would eventually become essential for looking at coral reefs through time, and for uniting the recent geologic past with the present. But it would take time for the importance of this idea to be recognized.

Toward the end of his life, James Hutton wrote a book to describe his theory. Perhaps because he was weak and ill when he was writing it, the book didn't capture much of an audience besides critics who didn't approve of a nonbiblical interpretation of geologic time.[17]

John Playfair, concerned that his friend's ideas were being dismissed, set out to write a book that communicated Hutton's theory more effectively than Hutton had done himself. He published the book, *Illustrations of the Huttonian Theory of the Earth*, in 1802, bringing the idea of uniformitarianism to a broad audience with real-world examples.[18]

Geologist Sir Charles Lyell was so inspired by John Playfair's book that he set out to apply the idea of uniformitarianism to explain geologic examples from around the world. In his own book, *Principles of Geology*, first published in three parts between 1830 and 1833, Charles Lyell popularized the phrase that is today synonymous with uniformitarianism: the present is the key to the past.[19] Mary Lyell, Charles's wife and a snail researcher, was likely a contributor to Charles's work and the many editions of *Principles of Geology* published during the nineteenth century.[20]

*Principles of Geology* explores how features like mountains, deltas, volcanoes, veins of granite, coal, and sandstone formed by processes that are still at work.[21] *Principles of Geology* made the case that we live in a world of small, incremental changes. Over geologic time, these small changes add up. Given enough time, small changes can form mountains and also reduce them to sand.

When Charles Darwin boarded the HMS *Beagle* in 1831 to start his epic voyage, the 22-year-old had no idea how much *Principles of Geology* would influence him, that its author would eventually become his mentor, or that the trip around the world would have a recurring theme of gradual changes adding up to monumental effects. The voyage would fuel his research for the rest of his life. He made observations in the Galápagos that would lead, decades later, to the theory of natural selection, the groundwork for our modern understanding of evolution (we'll explore how it relates to coral in chapter 9). But long before his natural selection work, Darwin discovered how atolls—odd, doughnut-shaped coral reefs—form in the Pacific and Indian Oceans and, more broadly, how reefs form and grow.[22] The principle of uniformitarianism helped him figure this out.

A trip can be colored by the book you bring along. Pack a horror novel and you may spend what would have been a relaxing beach vacation hiding under your beach towel. Charles Darwin had many books on board the *Beagle* that would influence him along the way. When the ship left England, the first volume of *Principles of Geology* was on board. Over the five-year voyage, Darwin was sent the other two volumes as they were published.[23] The books likely encouraged him to apply uniformitarianism to the curious things he was witnessing at sea and at ports of call, leaning on processes he could see happening in the present in order to understand how coral atolls had formed in the past.

In the nineteenth century, many European oceangoing expeditions were motivated more by scientific curiosity than by conquest as they had been previously, but it was still a time of imperialism with Western countries reaching their tentacles into other areas of the world, often with vast and negative impacts on the communities and cultures they encountered. Charles Darwin's account of the trip, the travelogue *The Voyage of the* Beagle, can seem shocking today, particularly his observations from countries with slavery and places where Christian missionaries disregarded local culture, but the majority of his observations were of the natural world instead of the human world. Even though this voyage was on a ship, most of his observations were made on land, perhaps because that's where he found geology and biology that intrigued him and perhaps also because he was prone to seasickness. A few weeks after the *Beagle* left England, Darwin wrote in a letter to his father, "If it was not for sea-sickness the whole world would be sailors." The only food he could keep down at the time was raisins.[24]

～

Atolls, scattered throughout the Pacific and Indian Oceans, have long been a hazard for ships. A ship would be in the middle of the ocean—no land in sight for days—and all of a sudden the sea would shallow to mere feet, its color turning from navy to turquoise speckled white with crashing waves. There, right below the surface, would be a reef full of colorful corals and tropical fish.

It was a mystery how atolls could form in the deep water of the open ocean even though reef-building corals needed to live in shallow water in order to have enough sunlight to grow. The prevailing theory in the early nineteenth century, which Charles Lyell had included in *Principles of Geology*, was that corals grew on the cratered summits of undersea volcanoes, mountains that happened to stop growing when their summits were at a shallow depth where corals could flourish.[25] The captain of the *Beagle*, Robert FitzRoy, had been tasked with investigating this theory on the voyage.[26] Charles Darwin was unconvinced. What were the chances that hundreds of undersea mountains would all happen to grow to the same height, with their summits at just the right depth for coral to grow?

One of the most intriguing aspects of these coral reefs in the middle of nowhere is their circular shape. Some atolls have a circle of low islands above water. Others have no islands, just a circle of reef below the surface. Some are huge, stretching across hundreds of square miles. Typically, only a small fraction of that area is land above sea level. In the Maldives, many atolls are clustered together. Other atolls are single, isolated reef circles surrounded by the deep sea.

Charles Darwin got an in-depth look at an atoll when the *Beagle* stopped at the Cocos Keeling Islands in the Indian Ocean late in the voyage, but he set the groundwork for his own theory about atolls much earlier in the trip by observing other geologic processes.[27]

Uniformitarianism is like a puzzle. To solve it requires fitting a piece of the past together with a piece of the present. James Hutton had witnessed present-day erosion and deposition as a farmer. He found the matching puzzle piece in the rocks at Siccar Point. Charles Darwin had one puzzle piece—atolls. He found a matching puzzle piece that helped explain how they formed on the coast of Chile, over 3,000 miles from the nearest atoll.

In February 1835, when Darwin and the *Beagle* were in Chile, a huge earthquake uplifted land along the coast. According to local residents, the land around the Bay of Concepción had been raised two to three feet. About 30 miles southwest of Concepción on Isla Santa María, there was much more uplift.[28]

"Captain Fitz Roy found beds of putrid mussel-shells *still adhering to the rocks*, ten feet above high-water mark," Charles Darwin wrote (emphasis his own) about the uplift of Isla Santa María.[29] Locals remarked that in the past, they dove to collect these mussels from the shallow seafloor. Suddenly, the shells and seafloor were—on land. There had been earthquakes in the past, according to locals, so he reasoned that there might be evidence upslope that the seafloor had been uplifted before. His suspicions were correct. He walked uphill, finding seashells at least 600 feet above sea level. His idea—that uplift happened in the past much as it happens in the present—applied uniformitarianism. With the earthquake, he had experienced the mechanism for uplift, when the Earth's surface rises. He didn't know that it was caused by the movement of tectonic plates, but he could see how small changes—a few feet of uplift every so often—could add up over time to the height of the Andes.[30]

The Chilean earthquake and uplifted land helped him realize that if land could be lifted up, then it must also sink down. He reasoned that to stay in balance, the Pacific could be sinking down (which is called subsidence) as South America lifts up. Years later, in his book explaining coral reefs, he would write, "And when we consider how many parts of the surface of the globe have been elevated within recent geologic periods, we must admit that there have been subsidences on a corresponding scale, for otherwise the whole globe would have swollen."[31] Subsidence, he would come to understand, was the key to forming coral atolls, because as land sinks downward, reefs grow upward.

The *Beagle* left South America and traversed the Pacific, navigating around atolls. It didn't make any stops, so Darwin's first glimpse of atolls was the maritime equivalent of a "drive-by." When the ship passed near an atoll, Darwin would climb the mast to get a better view of the anomalously shallow water and tiny islets in the middle of the open ocean. The atolls seemed completely out of place to him.[32]

After a month of sailing across the Pacific, the *Beagle* reached Tahiti, where Darwin hired local guides to lead him on a mountain hike. While hiking to 2,000 or 3,000 feet above sea level and marveling at how the island's plants formed zones with altitude, creating concentric circles of flora, he got a bird's-eye view of the nearby island of Mo'orea.[33] It was mountainous yet smaller than Tahiti, sitting in a circle of light blue, a shallow platform shelf, within a circle of waves crashing at a shallow reef. Beyond the crashing waves was the dark blue of the open ocean. Just like the plant life, the underwater realm had concentric circles. He compared his bird's-eye view of Mo'orea to a framed picture: the island as the picture, the light blue shallows as the mat, and the ring of coral with breaking waves as the frame.[34]

High above sea level on the flank of a Tahitian mountain, he had a eureka moment about what was happening below sea level. He found another key piece of the uniformitarianism puzzle, noticing that the circle of reef around Mo'orea was the same shape as an atoll. Mo'orea, he speculated, was a mountain sinking out of sight. What was happening in the present at Mo'orea, he reasoned, had happened in the past to form atolls. On the coast of Chile, he had seen how uplift sent marine life and the seafloor onto land, but in this case the seafloor was sinking down. He knew that living corals needed to be at sunlit depths, and so, applying uniformitarianism, he reasoned that corals in the past must have lived at sunlit depths, too. If coral reefs formed at the fringes of islands in the middle of the ocean and then the islands sank out of sight, the reef could remain alive if the corals kept growing upward as the land sank downward. The living corals would need to build upon previous generations of coral in order to stay in the shallow water. Darwin argued that the constant expansion of corals upward over time could produce limestone hundreds of feet thick.[35] We now know, thanks to rock cores drilled at atolls, that it can actually produce limestone that's several thousand feet thick, deeper than the tallest skyscrapers are high.

Back aboard the *Beagle*, Darwin used nautical charts and his observations to categorize the world's tropical islands into three groups. Some had a large, high mountain at their center and a reef fringing the shore. Others had a smaller mountain, a barrier reef a short distance from shore, and a calm lagoon between the barrier reef and the island. And atolls had a circle of reef and lagoon and no central mountainous island.[36]

With these three types of tropical islands, Darwin saw atoll formation as a three-step process. Step 1 looks like the large island with a fringing reef, step 2 looks like the smaller island with a barrier reef some distance from shore, and step 3 looks like an atoll. Transitions between these three stages were gradual, Darwin proposed. Over time, land sank downward (i.e., subsided) while corals grew upward.

"We see in each barrier-reef a proof that the land has there subsided, and in each atoll a monument over an island now lost,"[37] Darwin concluded, marveling at the scale of these features. "We feel surprise when travellers tell us of the vast dimensions of the Pyramids and other great ruins, but how utterly insignificant are the greatest of these, when compared to these mountains of stone accumulated by the agency of various minute and tender animals!"[38]

It took a great deal of time for those minute and tender animals (a.k.a. corals) to grow upward as the land sank downward. Much as Hutton had a sense of the amount of time needed to form the rocks at Siccar Point even if he had no way of dating the rocks, Darwin had a sense of the amount of time needed to form coral atolls.

By April 1836, when the *Beagle* sailed into the lagoon at the center of the southern atoll of the Cocos Keeling Islands (which were called the Keeling Islands at the time), he had developed his theory that atolls form at the site of subsiding volcanoes, and he was keen to apply his theory to a real atoll. "Considering that Keeling atoll, like other coral formations, has been entirely formed by the growth of organic beings, and the accumulation of their detritus, one is naturally led to enquire, how long it has continued, and how long it is likely to continue in its present state," he would later write in *The Structure and Distribution of*

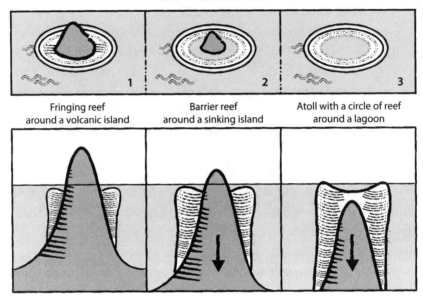

**How to make an atoll**

1. Fringing reef around a volcanic island
2. Barrier reef around a sinking island
3. Atoll with a circle of reef around a lagoon

*Coral Reefs*.[39] His book on coral reefs offers an explanation of all types of reefs but has a particular focus on atolls.[40] I find Darwin's quote striking because nearly two centuries later we still also wonder how long coral reefs will continue in their present state, not out of general curiosity but instead because of the numerous threats they face in the Anthropocene.

The Cocos Keeling Islands sit in the Indian Ocean about 800 miles southwest of Jakarta, Indonesia, and 1,300 miles from northwest Australia. Today they are a territory of Australia with about 600 residents. Most are Cocos Malay people, descendants of those recruited from South Asian countries in the 1800s to work on plantations run by the British.[41] Charles Darwin explains in *The Voyage of the* Beagle that several years before the *Beagle* arrived, the slaves on one plantation ran away to join the other plantation where workers were paid.[42] They freed themselves, but they would come to find that their wages were paid in a currency that had value only at the plantation's company store.[43] "The Malays are now nominally in a state of freedom, and certainly are so, as far as regards their personal treatment," he wrote, "but in most other points they are considered as slaves."[44]

Twenty-six islands are arranged in a horseshoe shape on one atoll forming the South Keeling Islands. One island—North Keeling—is far from the others on a small, second atoll. A NASA satellite image shows the South Keeling Islands—some only about as wide as a soccer field—and light swirls of sand with darker patches of reef in the lagoon within the horseshoe shape.[45] Closer to the center of the lagoon, the blue color deepens with water depth. At the edge of the atoll, the light blue of the shallows turns to dark blue within a few pixels on the image, a color change not unlike what one sees on the steep-sided limestone platforms in the Bahamas. In a way they are similar. In both cases, limestone is growing up as the Earth's crust sinks down (although Cocos Keeling atolls are sinking much more rapidly than Bahamian platforms). At an atoll, that layer cake of limestone sits atop a subsiding mountain, typically a volcano.

Because the coral construction project is ongoing at an atoll, Darwin could study corals in the present to learn how atolls formed in the past. To figure out how deep corals can grow, he and Captain FitzRoy took depth soundings around the Cocos Keeling Islands.[46] At the time, taking a depth sounding meant using a sounding line (also called a lead line), which has a lead weight, often shaped somewhat like a small butternut squash, at the end of a very long rope marked with its length. The weight is dropped overboard, eventually reaching the seafloor, and the depth is recorded as the amount of rope between the lead weight and the ship. At the slightly wider end of the metal squash, a hollowed-out base would be filled with something sticky like tallow (hardened animal fat) to pick up a sample of sand or mud from the seafloor where it landed.

"It was found that within ten fathoms [60 feet or 18 meters], the prepared tallow at the bottom of the lead, invariably came up marked with the impression of living corals, but as perfectly clean as if it had been dropped on a carpet of turf," wrote Darwin in *The Voyage of the* Beagle.[47] The tallow made contact with coral, not sediment, on the seafloor and so did not bring anything up with it besides the impression of the poor corals that were smacked with animal fat on a metal squash.

Slightly farther from the islands, Captain FitzRoy found corals to a depth of 120 to 180 feet (37–55 m or 20–30 fathoms). When the seafloor was deeper than that, the tallow collected only sediment, not impressions of corals. Given that present-day corals were constructing reefs in shallower water, corals in the past must have done the same, Darwin suggested. He concluded that "from the fact of the reef-building corals not living at great depths, it is absolutely certain that throughout these vast areas, wherever there is now an atoll, a foundation must have originally existed within a depth of from 20 to 30 fathoms from the surface." (We now know that some corals can live in deeper water with less light, but Darwin and FitzRoy found no evidence of that.)

He made a point of how unlikely the theories of the day (including the one in Lyell's book) were, given the very specific depth requirements of corals:

> It is improbable in the highest degree that broad, lofty, isolated, steep-sided banks of sediment, arranged in groups and lines hundreds of leagues in length, could have been deposited in the central and profoundest parts of the Pacific and Indian Oceans.... It is equally improbable that the elevatory forces should have uplifted throughout the above vast areas, innumerable great rocky banks within 20 or 30 fathoms, or 120 to 180 feet, of the surface of the sea, and not one single point above that level.[48]

In order to measure the depth of the ocean beyond the shallow water of the Cocos Keeling Islands, Captain FitzRoy took another depth sounding about a mile from shore. The sounding line, 7,200 feet (about 2,200 m) long, was not long enough to reach the bottom. The sides of the island plunged into unfathomable depths.[49]

When he wasn't making sense of the soundings, Darwin was on the top of the atoll getting to know the corals as best he could in that prescuba era. He waded in the lagoon at low tide and used a pole to leap across breaking waves at the outer edge of the atoll to see the corals in the surf.[50] He observed that the coral species that lived in the calm water were different from those in the high-energy waves—a pattern he learned about from his Indigenous guides months before while peering down at corals from a canoe near Tahiti[51] (and the basis of reef zonation, which we'll explore in the next chapter). He described corals in different parts of a reef as "probably all adapted to the stations they occupy."[52]

While he was adamant that the foundations of the coral atolls must be subsiding, he couldn't find much direct evidence to support this idea. The direct evidence had sunk out of sight. But he did find indirect evidence. He observed that old coconut trees were undermined and falling. He learned from local residents that a shed with its footings in the intertidal zone used to be above the high tide line. He asked about past earthquakes and learned that there had been three in the past decade, one severe.[53] This all supported the idea of sinking, but it was not direct evidence. The land had sunk below where science at the time could detect it. Today there are methods to monitor subsiding land. At that time, scientific instruments and sensors for field research were often no more high-tech than Captain FitzRoy's sounding line.

Some stories of science discovery through history feature researchers proving each other wrong. They are all about arguing. When Charles Darwin would eventually share his theory about coral reefs on sinking mountains, he would face some of this response. But the story of uniformitarianism that connects James Hutton, John Playfair, Charles Lyell, and Charles Darwin is about building on each other's ideas. Yes, Charles Darwin's theory of coral reefs did render other theories, including Charles Lyell's, incorrect, and these four people had other influences too, but in this case, they learned about an idea and then added their own spin to it. It's the philosophy that improv actors use—a "yes, and" approach—applied to science. John Playfair learned James Hutton's idea of uniformitarianism and was excited to share it with a wide audience. Charles Lyell learned from Playfair's book and was inspired to apply uniformitarianism to examples all over the world. Then Charles Darwin, inspired by Lyell's work, used the theory to explain coral atolls.

The story also contains an excellent example of how we should all react when our own theory of coral reef formation is replaced by a new and better theory. Three months after the *Beagle* returned to port, Charles Darwin shared his new theory over lunch at Charles and Mary Lyell's house.[54] I wonder whether he felt a little trepidation as he headed to the Lyells'. If his theory was correct, then Charles Lyell's theory of atoll formation was wrong. What's the etiquette related to quashing your luncheon host's theory of reef formation? At lunch, Darwin first shared his research on the uplift of the Andes, possibly because it supported his host's ideas and was bound to make everyone happy, but he eventually also shared his new theory about atolls.[55] It would have been human nature for Charles Lyell to be skeptical or grumpy since this new theory would mean that the one he had published in his book was incorrect. But no, he was positively giddy when he learned the idea of sinking islands at coral atolls. The new theory applied uniformitarianism ideas perfectly—giant reef structures forming because of an accumulation of gradual changes. At this nineteenth-century Victorian lunch, Charles Lyell was purportedly shouting, laughing, and dancing around the room.[56] He was overjoyed that someone had applied the ideas he advocated in *Principles of Geology*, even if it meant that his book would need an update. "My whole theory is knocked in the head," he later wrote to a colleague.[57]

After the lunch, he wrote a letter to Darwin declaring, "I could think of nothing for days after your lesson on coral reefs, but of the tops of submerged continents." But he continued with a cautionary tone. "It is all true, but do not flatter yourself that you will be believed, till you are growing bald, like me with hard work, & vexation at the incredulity of the world."[58]

Three years after he published *The Voyage of the* Beagle describing his whole adventure, Charles Darwin published *The Structure and Distribution of Coral Reefs* to more completely describe his reef theory.

The new theory of coral reef formation had supporters and critics. Darwin's idea of sinking islands may have described atoll formation well, but not all reefs were on sinking volcanoes. Some critics may have been spurred to propose alternative theories of reef formation because Darwin made bold declarations that invited skepticism (for example, writing, "I venture to defy any one to explain in any other manner how it is possible"[59]). Several of the alternative ideas proposed in the wake of Darwin's theory are well described in David Dobbs's intriguing book *Reef Madness: Charles Darwin, Alexander Agassiz, and the Meaning of Coral*.[60] This experience taught Darwin that proposing a theory without much evidence can lead to conflict, and he disliked conflict.[61] His theory and the arguments with it fueled a century of reef research and spawned numerous other theories.

"If we regard the question of the formation of the foundations of coral reefs honestly, we are forced to admit that all our theories and considerations are mere camouflage for our lack of knowledge," John Stanley Gardiner wrote in his 1931 book *Coral Reefs and Atolls*, 95 years after Darwin had developed his theory of sinking undersea mountains.[62] Stanley, as he was called, was a zoologist who studied reefs in the Indian and Pacific Oceans, including atolls (and he was also, for full disclosure, my great-great-uncle—a fact I learned recently that makes me wonder whether there could be a gene that codes for interest in coral reefs). In Stanley's time there was a lack of data about what was at the roots of atolls, but the last sentence in his book turned out to be prophetic: "The time is not far distant when the surveyors will chart not only the ocean floor in detail, but also the depths—and perhaps the nature—of the rocks underlying our coral atolls."[63]

With time, and advancements in technology, we've learned that Charles Darwin was right about subsiding mountains as the foundation of atolls. Atolls are isolated limestone platforms that sit atop undersea mountain pedestals. We've also learned how Earth's tectonic plates fit together, which explains why some areas of the ocean, especially in the Pacific, are spotted with volcanoes that become the foundations of atolls when they subside.

Deep cores drilled on Enewetak Atoll in the Marshall Islands in 1952 found igneous rock below more than 4,000 feet of ancient to recent coral reefs,[64] far deeper, and reflecting far more subsidence, than Darwin had speculated. This discovery confirmed that atoll limestone was forming on a sinking volcano. The core wasn't drilled specifically to test the theory of atoll formation. Rather, the Enewatak core drilled by the US Geological Survey[65] was part of the preparations for nuclear bomb tests during the Cold War, which severely damaged Enewetak and Bikini Atolls and spread fallout throughout the Marshall Islands, harming wildlife and human communities.[66] Someone eventually added a handmade sign next to the borehole proclaiming, "Darwin was right." He was indeed right about the foundation rocks, but there was more to learn about other factors affecting the growth of atolls.

By the early twentieth century, the effect of ice age cycles on global sea level and reefs was finally starting to be understood.[67] We now know that sea level has been swinging up and down while the mountains below atolls have sunk—two processes happening at once, which adds an extra layer of complexity to the formation of each atoll. In a 2013 study, scientists modeled that complexity by combining the effects of volcano subsidence with sea level fluctuations over the past 400,000 years. The model results showed that a reef's shape depends on the rate at which the corals grow, the rate of subsidence, and the rises and falls of sea level.[68]

On sinking atolls, fossil reefs are found deep underground, below layers of coral and sediment from more recent times. A core taken from the Cocos Keeling Islands found a fossil reef deep within the limestone, 26 to 36 feet (8 to 11 m) below current sea level.[69] Age dates show that it formed during the last interglacial—the same time as the reefs that are now fossilized at the fringes of islands on more stable platforms like the fossil reefs on San Salvador and other islands throughout the Caribbean. Charles Darwin recognized that there were many relatively young fossil reefs on the edges of tropical islands, but because sea level changes due to ice age cycles were unknown at the time, he concluded that all fossil reefs on land must have been uplifted,[70] much like the seafloor and putrid mussels that he saw uplifted by the earthquake at Isla Santa María, Chile. Because fossil reefs at the edges of islands are common in the Caribbean, he assumed that the whole Caribbean region had been uplifted recently. But in fact, it was only sea level changes that put many of those fossil reefs on land, not uplift: the slightly higher sea level during the last interglacial period meant that, for tectonically stable islands like those in the Bahamas, what was once the shallow seafloor is now land, and the corals in many cases grew right where they fossilized when that area was covered in water.

That brings us back to the isolated limestone platform where we started, San Salvador, and its largest fossil reef, which is where we're headed next. Today, it's possible to walk across the fossil remains of a San Salvador coral reef that thrived about 125,000 years ago. On other islands that have been uplifting, in the Caribbean and beyond, it's possible to walk across fossil reefs of several different ages. In the next chapter, we'll take a closer look at these now land-bound reefs and, using uniformitarianism, see how present reef ecology has been the key to interpreting their past.

# CHAPTER 3

# REEFS AT THE SHALLOW END OF DEEP TIME

Each morning at the Gerace Research Centre on San Salvador, researchers and classes load up their trucks with supplies and battered coolers of lunch foods, including loaves of the bread baked on-site and jars of peanut butter, and head into the field. In my case, "the field" often meant a fossil coral reef.

The Gerace Research Centre has a collection of sturdy old trucks painted Bahamian flag blue that researchers use to get around San Salvador. Each truck is given a letter of the alphabet as a name. My favorite—Truck E—was a bit smaller than the rest of the trucks. I could reach the pedals when the bench seat was all the way forward, and I could see over the steering wheel as long as I sat on a pillow.

It's hard to get lost on an island with one main road, especially if that road basically makes a circle around the perimeter of the island. Driving an old pickup truck around San Salvador takes some getting used to, at least it does when most of your past driving experience is, as mine was, in small hatchbacks. There's even more of a learning curve when driving on what Americans consider to be the wrong side of the road, especially when the truck has the steering wheel on the left, which is the wrong side of the vehicle if you are driving on the left. It made me feel like I was driving on the shoulder. With the window down, branches would whip my left shoulder as I tried to keep to the correct (left) side of the road.

There are at least 13 fossil reefs on San Salvador, all of which formed during the last interglacial period when sea level was about 20 feet (6 m) higher than it is today.[1] Only three fossil reefs are well studied, all on the west side of the island. One of those three fossil reefs is embedded in the rocky headland at Sue Point,[2] which is just north of a bay where Club Med tourists bake in the sunshine. A second is a small fossil reef that sits in the stubby cliffs of the island's southwest corner blanketed by rocks made of ooid sands that once resided on the shallow seafloor.

The most well-studied fossil reef is also the island's largest. That's where we're heading in this chapter. We looked at this fossil reef in chapter 1 to understand what lived there—the corals, snails, clams, and oatmeal flakes of algae. Now we'll explore the ecosystem, applying uniformitarianism to learn how much it has in common with typical reef ecology and fossil reefs throughout the Caribbean.

It's a short drive south from the research center, in Cockburn Town[3] (which is pronounced "co-burn," except when tourists mispronounce it and provoke giggles and jokes about nudity and sunscreen). The largest town on the island, Cockburn Town is home to fewer than 300 residents, about a third of the island's population. The town's congregation of bright cement and limestone buildings—pink, white, turquoise, orange, and yellow—is a place where you might find dogs, goats, and chickens wandering tropical gardens or lounging on the porches of colorful homes. When school gets out for the day, the town fills with children playing and laughing in crisp navy and white uniforms. The town is charming, but for paleontologists and geologists, Cockburn Town's biggest attraction is the fossil reef.

I spent weeks identifying thousands of clams and snails between the corals in this fossil reef when I was working on my PhD dissertation,[4] and I learned that every species that I could find on San Salvador's beaches, which I combed regularly, was also in the fossil reef. To avoid hammering the fossils out of the rocks, which would disturb the outcrop for future scientists, I would identify the mollusk species in place and count their numbers within a grid of squares to get a sense of how they were distributed in the fossil reef. I would check species identifications with field guides to modern shells, the types used by shell collectors. The field guides had fantastic photos showing the most colorful versions of the shell of each species. In contrast, their fossilized counterparts were dull white to concrete gray, the colors having leached away over time, and many shells were just fragments of what they used to be. With the book open next to my squares, I could see in one glance what a snail was like in the present and the past, a clear case of uniformitarianism in action.

If you were to head to the Cockburn Town fossil reef, you would park your blue truck on the west side of town where a giant iguana statue stands sentry. Walk north from the remains of an old dock, a rusty ruin, along the edge of the turquoise ocean and scramble over the rocky coast. At some point you would start seeing abstract curved shapes in the gray rock. Some very clearly look like the skeletons of reef corals. Others are harder to identify. The long-dead inhabitants of a coral reef are underfoot.

Eventually you would come to a broad, open area. This used to be the center of a quarry carved into the rock, which is why the fossil reef is so well exposed in this location. On the west side of the quarry, limestone meets the turquoise sea and waves crash on the wall of rock. On the east side, a stand of skinny trees divides the quarry from the town. On limestone islands, quarries like this were historically an important source of building stone, which implies that fossils fill the walls of Cockburn Town's older buildings.

This place is a time machine. Along the coast, over 2,100 feet (650 m) of fossil coral reef is exposed,[5] giving a glimpse of what coral reefs were like before the Anthropocene, in a world without human-caused problems like pollution and climate change. This reef formed in the late Pleistocene, a time when modern humans were still a relatively new species.

Most of the rocks exposed on the surface of San Salvador are Pleistocene in age, including the Cockburn Town fossil reef. The Pleistocene stretches from about 12,000 years ago back to 2.58 million years ago, but the Pleistocene rocks on San Salvador are at the younger end of that range. Around the fringes of the island are younger Holocene rocks that formed over the past 12,000 years.[6] The oldest rocks on the island, at least 220,000 years old, are called the Owl's Hole Formation.[7] In geologic terms, a "formation" is a layer of rock. The Owl's Hole Formation is named after a hole in the limestone, about 33 feet (10 m) deep, in the southwest corner of San Salvador in which these oldest rocks are found. Because younger layers of rock are above older on this layer cake of limestone, the oldest rocks are at the bottom of the hole, and yet they are still quite young compared with the nearly 4.6 billion years of Earth history. Legend has it that Owl's Hole was named after a snowy owl that once lived in the tree that grows from the bottom of the hole and whose treetop is not far above eye level when one is standing at the surface. I'm not able to find a recorded observation of a snowy owl—an Arctic species—this far south, so the story may not be true, but I imagine that many birds have called the tree, and the hole, home.

The Owl's Hole Formation rocks are capped by a fossil soil (called a paleosol), a rusty-red layer indicating a time when sea level was low and the area was land during a glacial period. Above the paleosol is the Grotto Beach Formation—a layer of rock formed from the seafloor at the top of the platform when sea level was high during the last interglacial between about 130,000 and 115,000 years ago. Some of this rock is made of ooids. Some of it is made of fossil coral reef, like the abandoned quarry near Cockburn Town.[8]

The Cockburn Town fossil reef was studied in the late 1970s by paleontologist Al Curran and geologist Brian White, the pair of Smith College professors who introduced me and countless other college students to San Salvador's geology and paleontology over several decades. With some of those students, they created a geologic map of the reef and surrounding limestone in the 1980s.[9] Since then, the fossil reef has been the focus of numerous research studies including some of my own. This might be one of the most scrutinized fossil reefs in the world.

Paleoecology is the study of the organisms that lived in an ecosystem based on their fossil remains,[10] which is like analyzing the crime scene of this reef. Al, Brian, and their team of students set out to analyze that crime scene. They turned to what's known about modern reef ecology, using the present as the key to the past (the principle of uniformitarianism described in the previous chapter) to understand the ecology of the fossil reef based on the corals preserved within it.

As with modern reefs, corals created the structure of the fossil reef that became a habitat for other life much in the same way as the trees of a forest. Because corals tend to be fossilized where they lived, they can tell detailed stories of a reef, revealing its ecological patterns.

Nearly all the coral species in this fossil reef are still alive today, so Al and Brian referred to a field guide, *The Peterson Field Guide to Coral Reefs*, to identify this fossil reef as a bank/barrier reef,[11] which isn't as far offshore as a barrier reef and resembles a fringing reef that is not connected to land. At its peak, sea level was about 20 feet (6 m) higher when this fossil reef formed.[12]

Overall, they found the same broad-scale pattern of coral species in the fossil reef at Cockburn Town as is typically found in healthy living reefs. Fossil reefs throughout the Caribbean and beyond, including some older and younger fossil reefs, usually have the same pattern too, as we'll discover later in this chapter, implying that reefs have been building in the same way over hundreds of thousands of years.

∼

Coral reefs often have a pattern of zones much in the same way that cities often have a pattern to their built environments. In a city, a central business district is usually surrounded by mixed-use neighborhoods surrounded by suburbs and, even farther away, exurbs or farmland. Similarly, a reef has a suite of distinct zones oriented as bands that parallel the coast.[13] The zones can look a bit different depending on whether you are looking at a fringing reef, a barrier reef, or an atoll, and the coral species found in the zones vary depending on where you are in the world, but the general patterns can still be seen. This pattern of zones is often disrupted in modern struggling reefs, but it's still helpful to understand since such a pattern indicates a healthy reef, whether it's in fossils from the geologic past or in living reefs underwater.

If you were to snorkel from a Caribbean shore fringed by an archetypal and undisturbed coral reef, you would see the pattern. First, you would encounter a calm-water lagoon, also known as the backreef zone. The water would gradually deepen, perhaps to 20 or 30 feet (6–9 m), as you kicked your flippered feet. If you looked through your mask at the seafloor, you might see patches of reef—called patch reefs—with large heads of star coral (*Orbicella*) and finger coral (*Porites porites*), which can look like carpets of lumpy fingers. Thickets of delicate branching staghorn coral (*Acropora cervicornis*) would be found in the backreef too, along with other seafloor neighborhoods like patches of sand, seagrass meadows, and collections of rocky rubble formed from the skeletons of dead corals.

If you kept kicking your fins, you would find that, counterintuitively, the water shallows as you get farther from shore because corals are building upward. Waves crash and sunlight permeates without the blue cast of deeper water. The shallowest area is the reef crest. Waves would start to toss you about as elkhorn corals (the other species of *Acropora* in the Caribbean, *Acropora palmata*) come into view, growing up near the sea surface. Elkhorn coral thrives in the reef crest's high-energy environment.[14] It prefers the extra sunlight in shallow water, and because its branches are sturdy, waves rarely cause colonies to break, but when they do, the pieces that are broken off are sometimes able to attach to the rocky seafloor, founding new colonies.[15]

A colony of elkhorn coral looks more like the antlers of a moose than an elk, with broad, flat areas at the end of sturdy trunks and branches (at least that's the case if you are from North America—the animal that is called an elk in Europe is called a moose in North America). Alive, elkhorn coral is yellow or orange thanks to zooxanthellae, the algae that live in its tissues. Like a moose prepping for a fight, the antler shapes are oriented to take on the waves. You might find that you couldn't snorkel across these antler shapes without painfully scraping your belly and killing coral. But if there were a spot where the coral colonies were a safe distance beneath you, you could swim across the reef crest. On the seaward side, you would typically find stronger waves than in the backreef, which is protected by the reef crest. If you looked back toward the reef

crest to get a view of the forest of elkhorn corals, below the canopy of flattened antlers, you would see fish darting and loitering between the coral trunks and branches.

If you looked down through your mask, seaward of the reef crest, you would see the seafloor falling away as the water deepened beneath you. This is the upper part of the forereef, which is where the highest diversity of coral species is typically found.[16] You would see less elkhorn coral as the water deepened and might find more staghorn coral interspersed with heads of star and brain coral. The corals might be along ridges perpendicular to the reef crest called buttresses. In even deeper water, you would find corals shaped like broad, flat plates.[17] Light is in short supply in deeper water, and the corals' flat shape maximizes the amount of sunshine that gets to their zooxanthellae for photosynthesis, much like the coral version of a solar panel. On some reefs, the forereef is steep, plunging rapidly into deep water at a wall on the side of the limestone platform. On others, the forereef slopes more gently. In either case, you would eventually get to a depth where sunshine couldn't penetrate, far beyond the reach of typical scuba equipment. The reef ends where it is too dark for symbiotic algae to survive.

While this pattern of zones running parallel to the shore is a good way to understand an average reef structure, each reef is unique. Reefs rarely follow the model precisely, just as individual cities rarely follow the pattern of built environments precisely. Ecosystems have patterns, but not precision. Each coast has its own unique topography, and this affects which areas of the seafloor are shallow, which areas are deeper, and where corals can live.

The amount of wave energy differs from reef to reef, too, and this affects which corals will thrive in each reef zone. Swiss scientist Jörn Geister defined six types of Caribbean reefs based on the amount of wave energy. A key difference between these six reef types is which coral species live in the reef crest, as each coral species has its own version of what is "just right" in terms of wave energy.[18] If there are too many waves, there won't be many corals in the reef crest; the waves are too fierce for even elkhorn coral to survive. With wave energy that's high, but not too high, elkhorn coral dominates the reef crest. With slightly less wave energy, staghorn coral is on the reef crest instead of elkhorn. Even lower wave energy and you will find star and finger coral in the shallowest areas.[19]

It's not just Caribbean reefs that follow predictable patterns. Worldwide, coral reefs have similar zones even if the species differ. Most have a shallow reef crest. On the protected side of the reef crest is a backreef zone with corals that prefer calm water. This zone can be crowded with corals in very shallow water on some Indo-Pacific reefs (the region that includes the Indian and Pacific Oceans). Or it may have deeper water and sparse patches of reef. Seaward of the reef crest are corals that prefer some depth.

Today a diagram of reef zonation can be found in nearly every marine biology textbook, but this pattern of reef zones wasn't well documented by Western science until the early 1950s, when newly developed scuba technology made it possible for scientists to explore reefs like fish. Before scuba, the pattern of corals in the shallowest reef zones could be interpreted without any equipment, by peering over the side of a boat in calm water and looking down to the seafloor. That's how Tahitians taught Charles Darwin what they knew about reefs in 1835, as we saw in the previous chapter.

Two of the earliest science pioneers to explore the ecology of coral reefs after the dawn of scuba diving were married couple Thomas and Nora Goreau, who studied Jamaican coral reefs.[20] Thomas F. Goreau was teaching physiology as a professor at the medical school of the University of the West Indies.[21] Nora Goreau, the first Panamanian marine biologist, was teaching high school science at girls' schools in Jamaica. They taught during the week. Weekends were for coral reef research, largely without funding in the early years.[22]

Scuba was still so new that Thomas couldn't buy the equipment in Jamaica, so he made his own scuba setup. He would dive and collect samples and bring them to Nora, on shore, for analysis.[23] Underwater, Thomas noticed that different coral species were found in different reef zones.[24] Looking at how corals change with water depth, Thomas saw a row of elkhorn coral taking the brunt of the waves in the shallowest parts of the reef. As depth increased, a thicket of staghorn coral gave way to heads of star coral and brain coral. The number of branching corals decreased while the number of rotund, massive corals and plate corals increased. In even deeper water, the number of massive corals decreased as the number of plate corals increased. Below about 100 feet deep (30 m), he found that almost all corals were shaped like flat plates.[25]

～

The pattern of reef zones is also found in fossil reefs. In the Cockburn Town fossil reef, a large thicket of fossilized elkhorn coral is the key to reconstructing the ecosystem and its physical environment. My brain coral thinking spot is surrounded by elkhorn coral entombed in sediment. Some of the fossil elkhorn coral colonies are in the same position they might have been in life, oriented to take on ancient waves. Others are preserved as a jumble of broken branches the size and shape of human limbs, all now gray and lifeless. Elkhorn coral has very small corallites—the limestone towers in which polyps live—so its surface isn't covered with meandering lines like brain coral, which has elongated corallites shaped to fill its mazelike pathways. The corallites of elkhorn coral look like goosebumps on the skeleton's surface where they haven't eroded away.

Elkhorn coral is predictably found in the shallowest parts of healthy modern Caribbean reefs—the reef crest—so we assume that it also lived in the shallowest parts of reefs that grew during the last interglacial period.[26] Looking at the wall of elkhorn coral in the Cockburn Town fossil reef, we can imagine the ocean just above, and waves crashing over what is now the remains of a limestone quarry.

Today the sea laps at the west side of the Cockburn Town fossil reef, eroding rock and fossils. But with a little imagination, it's possible to see the past—the yellow-orange corals, the fish that used to dart between their trunks, and the waves that once crashed overhead. It really wasn't that long ago. Look at the ocean just offshore today and imagine it about 20 feet higher, the height of a full-grown male giraffe. A calm backreef zone would have been where the town is today. The forereef zone would have been on the seaward side.

There are fossil reefs all over the world that formed at the same time as the one in Cockburn Town, and they typically have the same pattern of zones. As our planet headed into an interglacial time about 135,000 years ago, ice melted, warming seawater expanded, and global sea level rose nearly 400 feet (120 m) in about 5,000 years, flooding the tops of limestone platforms and the coasts of continents.[27] After this burst of sea level rise, reefs formed, and they are now preserved on the edges of islands that aren't subsiding and some mainland coasts across the tropics.

South of San Salvador sits the Bahamian island of Great Inagua, where a fossil coral reef forms the rocky northwest coast.[28] Head through the Windward Passage, the waterway between Cuba and Haiti, which both have fossil reefs,[29] and you'll get to Jamaica, where fossil

Elkhorn coral (*Acropora palmata*) is characteristic of shallow reef crests on Caribbean reefs. Alive (top), its branching structure can tolerate high waves, and when fossilized (bottom), the space between branches is filled with sediment, broken coral branches, and other fossils. Top photo by author, bottom photo by Sally E. Walker.

reefs on the north coast follow the classic pattern of zonation that Thomas F. Goreau saw in the living reefs offshore.[30] In the Florida Keys, fossil reefs formed at this time appear a bit different from others because they lack elkhorn coral, which scientists think may have been either because wave energy was low or because the climate wasn't right for elkhorn corals to grow at this time in Florida.[31]

But not all fossil reefs are well preserved. Case in point: there likely used to be fossil corals in what is now an understated attraction on the island of Grand Cayman. A short walk from West Bay Beach at the north end of the island's resort-filled west coast is a cluster of jagged black pinnacles of limestone known as Hell. There's a Hell Post Office in case you want to send a postcard from Hell. And of course, Hell has a gift shop. The rock features at Hell, jagged dark spires of limestone, are phytokarst, also known as biokarst, where limestone is destroyed by tiny algae and other biological agents like bacteria and fungi that bore away at the rock, dissolving it bit by bit,[32] obliterating the limestone and, with it, any evidence of fossil corals.

A 15-minute walk to the west of Hell are fossils of the former reef crest—elkhorn, staghorn, and star corals in limestone that hasn't been weathered into spires as it has been at Hell.[33] For someone who loves fossils, seeing them eaten away is a bit hellish, no pun intended, even if it's a natural process.

Sometimes humans instead of microbes erode fossil reefs. Just south of Cancún, Mexico, developers of the Xcaret theme park carved through a large fossil reef to create the park's topography. They blasted away rock from the fossil reef, but on the upside, the cuts they made through the rock exposed sections of the reef. About 10 years after Xcaret opened, a team of researchers studied the rock exposures and identified distinct reef zones.[34] Geologic fieldwork in a theme park came with some unusual safety concerns. For example, there were great exposures of the fossil reef in a moat that was home to crocodiles. The moat had been dug in an effort to contain the theme park's captive jaguars, which couldn't cross it. The team lowered a ladder into the moat, climbed down, and documented the fossil reef on its walls while listening to the jaguars' roar and watching for crocodiles. They found large elkhorn corals in the fossil reef crest, as well as a reef front and a large backreef lagoon with fossil patch reefs that lived during the last interglacial period.[35]

In the southern Caribbean you'll find massive fossil reefs on the islands of Curaçao and Barbados that have the pattern of reef zones, but they also hold evidence about how coral reefs were built and rebuilt over a much longer stretch of geologic time. On both islands, land is uplifting thanks to plate tectonics, so multiple fossil reefs form stair-step terraces around each island as the shallow seafloor rises upward and becomes land.[36] At the same time that land has been uplifting, sea level has swung up and down. With sea level high, reefs would form in the shallow water surrounding these islands, and uplift would send the reefs onto dry land. A new reef would form in the shallow water, which again would be uplifted. It's like a coral reef escalator that moves up one step, and one reef, at a time.

Unlike a classic limestone platform that sinks gradually so that the youngest limestone is at the top, which Charles Darwin described at the Cocos Keeling Islands, an island that's uplifting winds up with the oldest limestone at its highest points. On Barbados, the youngest stairstep, which grew 82,000 years ago, is down near the coast. A step uphill has a fossil reef that formed 104,000 years ago, and another stair above that formed 125,000 years ago, the same age as fossil reefs at the edges of tectonically stable islands like San Salvador. Farther uphill are older reefs, which formed 195,000 and 220,000 years ago, and there are yet older steps above these.[37] The reefs studied have the pattern of zones with elkhorn coral filling the reef crests.

In the case of uplifting islands that preserve multiple periods of reef building, we can see that the same pattern formed in reefs over and over as sea level rose and fell, as global climate was swinging warmer and cooler.[38] Because of these uplifting islands, we know that it's not just reefs from 125,000 years ago that are similar to modern healthy reefs. We know that reefs that formed repeatedly over the past 220,000 years have the same pattern.

West of Barbados, the island of Curaçao has a similar combination of uplifting land and sea level change that has left five fossil reef terraces on the island.[39] The 125,000-year-old fossil reef crests on Curaçao show the pattern of reef crest species that's expected with different amounts of wave energy. The reef crests are filled with elkhorn coral on the windward, or exposed, side of the island where waves are large. On the leeward (sheltered) side of the island, fossil reef crests have less elkhorn and more massive (boulder-shaped) corals, like brain and star coral.[40] Waves can't be fossilized, but their effects on coral species in the reef crests were fossilized on Curaçao.

While some of the most well-studied Pleistocene coral reefs are in the Caribbean, there are also striking examples in other tropical locations, which all add up to a detailed record of coral reef ecosystems over geologic time. In the Indian Ocean, fossil reefs have been found clinging to Precambrian granite on islands in the Seychelles.[41] Fossil reefs have also been found in Indonesia and on some Pacific islands.

Particularly well-preserved fossil reefs on the Huon Peninsula of Papua New Guinea also have a pattern to the corals.[42] As in Barbados and Curaçao, multiple fossil reefs form stair-step terraces on the Huon Peninsula because of the combination of tectonic uplift and sea level change, but the rate of uplift differs between the locations. When John Pandolfi was researching the paleoecology of the fossil reef terraces of the Huon Peninsula, he followed trails through the rain forest to climb the mountainside terraces, walking over the skeletons of Pleistocene corals that are now over half a mile (1,000 m) above the ocean surface. The fossil reefs on upper terraces (the oldest reefs) are too covered in vegetation for research, so he focused on the corals in the lower stair-step terraces, which record nine periods of reef building between 125,000 and 30,000 years ago.[43] "They basically just look like a staircase going up the mountainside," John says. He and his team found that all nine steps in the staircase have similar zonation of coral species.[44] Reefs grew with a similar pattern of corals each time. Much as in the Caribbean, John found that coral reefs in Papua New Guinea had a remarkable ability to persist.

Large fossil reefs can also be found within terraces around the coast of the Red Sea, between the Arabian Peninsula and East Africa. There are fossil reefs of several different ages on the west coast of the Red Sea in Egypt. The youngest formed during the last interglacial period (late Pleistocene) and the oldest formed over 300,000 years ago.[45] Like fossil reefs elsewhere in the world, these Egyptian fossil reefs have zones much like those of modern reefs, and nearly all the coral species identified in the fossil reefs are today living in the Red Sea.[46]

South of the Egyptian fossil reefs, on the coast of Eritrea, a late Pleistocene fossil reef includes some particularly unusual artifacts: stone tools have been found between the corals and sediments.[47] It's a reminder that there were humans on the planet when these reefs were alive, just not nearly as many as there are today. The teardrop-shaped hand axes and blades made from volcanic glass were found lying between cobbles of coral and within beachrock. The researchers deduced that the numerous tools they found were used at this site, which was a beach and a shallow fringing reef filled with corals, sand, and mollusks. These tools showed no signs of being battered as they would if they had washed in from elsewhere. Much is unknown about the owners of these tools, but it was clear to the researchers that whoever made the tools used them, at least in part, to harvest seafood like oysters and crabs. This is some of the earliest evidence of our ancestors eating seafood, just as coastal communities today often rely on seafood, including marine life from reefs.[48]

∿

While Caribbean fossil reefs typically have the classic pattern of coral zonation, most modern Caribbean reefs are now very different from their predecessors. These include the Jamaican reefs where the Goreaus described reef zones.

Shortly after Thomas, Nora, and son Thomas J. Goreau authored a 1979 article in *Scientific American* that described the coral zones off Jamaica's northern coast—including heathy thickets of elkhorn coral in the shallow reef crest and vast thickets of staghorn coral nearby[49]—a number of unfortunate events caused major changes to these reefs. Hurricane Allen in 1980 caused catastrophic damage, transforming the reefs in a day,[50] and disease started to kill elkhorn and staghorn coral.[51]

Within a couple of years, another disease would wipe out a species of algae-eating sea urchin, which, combined with the effects of overfishing, meant that macroalgae, a type of seaweed also known as fleshy algae, could grow virtually unchecked on the reefs.[52] Unlike the tiny unicellular algae that live symbiotically within corals (zooxanthellae), macroalgae cover reef surfaces with a slimy carpet. These algae can outcompete corals for space on a reef when they're not being eaten by grazing reef life, leaving corals with less room to grow.[53] Eventually climate change started to have direct impacts—in 1987 mass bleaching was first reported on Jamaican reefs.[54] Corals exposed to heat expelled their zooxanthellae and turned white, becoming weakened, starved, and more susceptible to disease.

"It's a disaster everywhere you look," exclaimed Thomas J. Goreau upon seeing the Jamaican reefs about four decades after he coauthored the *Scientific American* article with his parents. This was in the 2020 documentary *Coral Ghosts*, in which he visited the Jamaican reefs that had been his childhood playground with his brother Peter.[55] "Corals all sick and unhealthy looking. And now, everything's pretty much covered by masses of fleshy algae," Thomas explained. "What you see is dead reefs that are standing."[56]

Marine scientist Inilek Wilmot, too young to have known healthy reefs in Jamaica, was interviewed in *Coral Ghosts* when he was the manager of Jamaica's Oracabessa Bay Fish Sanctuary. He noted, "You go out and look at the reef, you have to guess what it used to look like because you're seeing skeletons of different corals that used to be alive. You can only imagine what was there."[57]

What happened to the reefs off the north coast of Jamaica was not an isolated incident, although it was an extreme case. Scientists were watching reefs throughout the Caribbean fall apart in the last two to three decades of the twentieth century, although the specific timing varied from place to place and some locations remained healthier than others.[58]

There are many potential causes for Caribbean reef decline, some more obvious than others: human population growth, overfishing, pollution, climate warming, and pathogens—which this book will explore in chapter 5. Overall, what we see today in Caribbean reefs is a struggling ecosystem. We humans may have invented scuba equipment just in time to glimpse the pattern of corals in reef zones before that pattern was disrupted, although some Caribbean reef struggles started long before. This is what makes the fossil record so valuable. It's preserving the story of what the world was like before these problems. Pleistocene fossil reefs can help us understand what modern reefs should be and could be.

While coral reefs worldwide are vulnerable in the Anthropocene, nowhere else on Earth have coral reefs been so prone to collapse as in the Caribbean.[59] Reefs in the Indo-Pacific are experiencing damage, particularly mass coral bleaching, which has become increasingly common, but Indo-Pacific reefs have so far been more likely to recover from damage than Caribbean reefs.[60] Why have Caribbean reefs been less able to bounce back? The unique geography of the Caribbean might be the key.

The various oceans of the world—the Atlantic, Pacific, Arctic, Southern, and Indian—are actually different basins of one global ocean. They are all connected and the water has no allegiance to a particular place. It flows around the world in ocean currents. However, the Caribbean is like a cul-de-sac of the ocean. It's at a bit of a remove. Currents do flow from the Atlantic through the Caribbean, but it's still pretty isolated.

It wasn't always this way. Before the Isthmus of Panama formed about 3.5 million years ago, cutting off the sea from the Pacific to the west, ocean water moved freely from Pacific to Atlantic, and so did marine life.[61] Once the isthmus formed, the Caribbean was isolated. Initially, environmental conditions changed in ways that made life easier for coral reefs in the Caribbean.[62] But one to two million years after the isthmus formed, an extinction event wiped out over half the coral species in the Caribbean.[63] This is the same time the current ice age began, with its oscillations between cooler glacial and warmer interglacial times and changes in sea level (more about that in chapter 6). It's possible that environmental changes of the ice age caused the extinction and the Caribbean's isolation made that extinction worse.[64] The extinction left only about 70 coral species in the Caribbean[65] during the timeframe when the fossil and modern reefs described in this chapter lived. In comparison, there are 831 recognized species of coral worldwide, and

758 of those species are found in Indo-Pacific reefs.[66] Because of the isthmus, coral species from the Pacific and Indian Ocean basins can't find their way into the cul-de-sac.

With fewer coral species, Caribbean reefs have fewer species that fill specific functions on a reef. For example, the reef crest zone is characterized by elkhorn coral because this is the only species in the region that thrives in the shallow water where waves break. And when a disease affects elkhorn coral, the reef crest can be left with a thicket of dead coral that becomes covered with slimy algae and provides less habitat for other species in the ecosystem.

Much as the people working in an office recognize they're in trouble when the sole IT person has left for vacation and computers are on the fritz, scientists recognized that Caribbean reefs were in trouble several decades ago as the health of key species started to decline.[67] It's possible that because Caribbean reefs have fewer coral species, they have lower resilience.

Resilience is the ability to cope with problems like environmental change. It doesn't necessarily mean there are no problems or that disastrous conditions have no effect on a resilient ecosystem. A disaster could happen and have terrible consequences. Yet a more resilient ecosystem has more ability to bounce back than a less resilient ecosystem.

According to the hypothesis of functional redundancy, ecosystems in which species overlap in their functions are typically more resilient.[68] For example, a reef with vast thickets of one coral species may look healthy, but it is less resilient than a reef with dozens of coral species because if a disease, heat stress, or something else caused that one species to die, the entire reef would be gone. A reef with many types of coral could carry on even if one species died.

It's possible that higher diversity in Indo-Pacific reefs means there is more redundancy, which makes them more resilient. But there will be limits to what reefs can endure. Over the last two decades, warming ocean water has been testing the resilience of reefs worldwide, including in the Indo-Pacific,[69] and future warming could make it difficult or impossible for reefs to bounce back.[70] There's evidence that other threats, like declining water quality, have killed corals in the Indo-Pacific too.[71]

Yet the existence of fossil reefs with the same pattern of zones suggests that, worldwide, reefs have been able to persist over deep time as sea level has swung up and down during the Pleistocene, forming and re-forming in the same way until recently.[72] The same corals that were abundant in Caribbean reefs of the Pleistocene were also abundant in Caribbean reefs until the early 1980s, according to John Pandolfi and Jeremy Jackson, who compared the paleoecology of fossil coral reefs across the southern Caribbean.[73] But the present is no longer the key to the past. Since the 1980s, coral reefs have changed in ways that they never have in at least 220,000 years.[74]

~

Modern reefs just offshore from San Salvador's Cockburn Town fossil reef have been changing too. Drive a few minutes south of the fossil reef and you'll find Fernandez Bay, part of the turquoise stripe in the tropical Rothko painting where a shallow lagoon is peppered with patch reefs. This is the lee side of the island, so the water is usually calm. A reef crest with elkhorn coral can be found on the north side where there are more waves, but the placid water on the west side is home to corals that prefer peace and quiet. Beyond the turquoise stripe, you'll see the platform edge and the dark blue stripe of deep ocean.

I got to know this bay well between 1999 and 2005 while my research team was collecting shells from the backreef. We were using the present as the key to the past, figuring out how much modern shells get tossed around by currents and waves (quite a bit, as it turns out) to understand what may have happened to the fossils.

We would stop the truck on the side of the road, don scuba gear on the narrow beach, and start paddling out to sea, leaving the late Pleistocene behind and heading into the living world on the modern seafloor.

Underwater shell collecting doesn't require stooping like shell collecting on beaches. With scuba gear, you can hover just barely above the seafloor, your head a bit lower than your feet, and pluck shells from the bottom of the shallow sea for each sample.

After so many hours underwater focused on the seafloor of Fernandez Bay, I learned where there were patches of sand lined with ripples and meadows of seagrass swaying back and forth. Farther out was an area where sea fans and other soft corals were rooted in the sand. From there, I could see the skyline of Telephone Pole Reef. The closer to the reef I would get, the more fish traffic I would see and the more crunching, chewing sounds of parrotfish I would hear.

Based on what we know about reef zones in the Caribbean, you might expect to see star coral and staghorn coral in the relatively calm water of Telephone Pole Reef. But in and around the bulbous heads of living star coral, what I saw was a coral graveyard. A carpet of mostly dead finger coral was increasingly covered with fuzzy, bright green macroalgae like Chia Pets. The finger coral was atop the bones of dead staghorn coral. I still found lots of live corals in the reefs of Fernandez Bay (particularly massive species like star and brain coral)—but in many spots they were far outnumbered by the dead.

Telephone Pole Reef doesn't draw lots of tourists or get mentioned in travel guides. As reefs go, it's pretty average. But this reef has a unique distinction: it's been watched closely for decades by researchers staying at the Gerace Research Centre. It's been under scientific surveillance since the 1980s, which means that we have a record of how its corals have changed.

In the early 1980s, Telephone Pole Reef was called Staghorn Reef, named after its most abundant species: staghorn coral, which was very much alive at the time. Thickets of staghorn coral filled the areas between the mounds of star and brain coral. However, by the mid-1980s virtually all the staghorn coral had died.[75] The dead thickets collapsed into piles of coral rubble, and finger coral started colonizing atop the dead staghorn coral skeletons. By the early 1990s finger coral had become a dominant coral on Telephone Pole Reef.[76] Some of the colonies were more than a meter across, which is a lot of fingers.[77] It was healthy and growing fast, but it wouldn't stay that way. By the late 1990s the finger coral was in decline.[78] And by 2002 virtually all the larger colonies of finger coral were dead.[79]

In less than two decades, reefs across the Caribbean transformed, and they continue to transform. As on Telephone Pole Reef, staghorn corals were wiped out on other reefs. Elkhorn corals died en masse as well. Other coral species have become increasingly vulnerable and the percentage of living coral in reefs dropped.[80] It's not unusual for colonies that have lived for centuries to die abruptly, their quiet skeletons remaining in reefs, often covered with macroalgae, until they crumble. And now changes like these have started to occur in reefs beyond the Caribbean, too.

Why did these reef transformations happen? While coral disease was the smoking gun, explaining the death of the staghorn coral, there is also a long list of contributing factors, nearly all caused or exacerbated by humans (which we'll explore in chapter 5). Dramatic changes to the corals within reefs, like what happened at Telephone Pole Reef and at reefs across the Caribbean recently, aren't typical in the reef fossil record.[81] In fact, Caribbean fossil reefs hold scant evidence of ecological change. Overall, the story of these fossil reefs is of one of stability and consistency, with the same coral species building reefs over time. When we consider modern tumult in reefs, the present doesn't look like the past. The changes we are seeing in reefs today are, in many ways, without precedent.

Uniformitarianism has its limits in the Anthropocene, at least for coral reefs. The algae-covered and increasingly vulnerable reefs of today are not the key to the past. But perhaps the opposite is true: understanding how reefs lived in the distant past, and what happened to them in the recent past, can help us find ways to help reefs survive in the present and, I hope, have a future.

Over the next few chapters, we'll step back in time from the present to recent decades, to centuries ago, and eventually to many thousands of years ago to see how the stories of past reefs can help us pinpoint the problems and find answers that can help modern coral reefs. First, we'll head to the most recent scene of destruction: modern reefs, where there's evidence in the skeletons of dead corals that can tell us more about reefs and how they have been changing.

CHAPTER 4

# INTO THE DEATH ASSEMBLAGE

I was underwater filling a sample bag with dead coral rubble on a reef in the Florida Keys when the light dimmed. I froze. The water around those shallow reefs is usually clear and most sunlight filters through. Certain wavelengths don't make it—red and orange, for instance, are a bit muted—yet there's still enough light that the corals and fish can have shadows. I looked up to see what caused the eclipse at the sea surface.

A flotilla of snorkelers in bright yellow vests were bobbing above me, eyes wide behind their masks, hands waving frantically. I could hear voices muffled by snorkels. They knew the rule—never touch coral—and they were making sure that I knew it too. Yet there I was in a shallow pit at the bottom of the reef, plucking out coral branches and small rotund skeletons and putting them in a mesh bag.

I looked like a caricature of what not to do on a reef. But I was on official business, as a research diver assisting a project by collecting dead coral skeletons from modern reefs to figure out what stories they can tell. The project was funded by the National Oceanic and Atmospheric Administration (NOAA) and led by paleontologists Ben Greenstein and John Pandolfi (whom we met exploring Papua New Guinea's fossil reefs in the previous chapter). The dead corals I was collecting from modern reefs could be thought of as potential future fossils at the current end of geologic time. My goal was to learn from the contents of the sample bags about how skeletons are altered after coral animals die, and what the remains of the dead can tell us about past environments and ecosystems. Over two field expeditions in 1994 and 1995 our team surveyed shallow and deep reef zones. I was on the shallow reef crew, investigating patch reef and reef crest sites, which at the time were unusually healthy for reefs in the upper Keys.

The shallow reefs are about five miles east of Key Largo, Florida, far enough out to sea that the land of the Keys was just barely visible on the western horizon as the boat arrived at the field sites. The Florida Keys themselves are made of Pleistocene fossil coral reefs and other petrified seafloor environments like an ooid shoal.[1] Each day, the NOAA boat would leave a Key Largo dock along a cut through fossil limestone and race across the waves. The blue of the sea would darken as the water grew deeper and then lighten as we approached patch reefs and then the reef crest.

The reefs adjacent to the Keys and South Florida collectively form a barrier reef, a comma of reef over 300 miles (480 km) long that runs from north of Miami to the Dry Tortugas in the Gulf of Mexico (the third longest barrier reef in the world after Australia's Great Barrier Reef and the Mesoamerican Barrier Reef off the Caribbean coast of Central America).[2] Known as the Florida Reef Tract, the reef has the zones typical of Caribbean reefs, with over 6,000 patch reefs[3] in a broad backreef area between the reef crest and the Keys and a deeper forereef zone seaward of the reef crest. The shallow seafloor near the Keys also includes a patchwork of other environments—seagrass beds, mangroves, sand flats, and places where the seafloor is solid limestone, formed at an earlier time (which we'll learn more about in chapter 6).

The majority of the Florida Reef Tract is within the Florida Keys National Marine Sanctuary. Some of the northern reefs are in Biscayne National Park, not far from Miami. Some reefs near Key Largo in the northern Keys are within John Pennekamp Coral Reef State Park. The south end of the Florida Reef Tract is within Dry Tortugas National Park. All of these reefs, and the state of Florida, are on a limestone platform attached to the North American continent.

The shallow reefs we studied are in the upper Keys, the northernmost end of the Florida Reef Tract. Carysfort Reef, one of the four we studied, is where the snorkelers caught me red-handed. Both the reef and the nearby lighthouse, Carysfort Reef Light, are named for a British ship that ran aground here in 1770 en route to Jamaica.[4] It's no wonder that the ship ran aground. In some places, the water is only a few feet deep.

Carysfort Reef Light was built into the center of a sprawling quilt of reef crest corals with lagoon patch reef corals on the protected side.[5] The lighthouse footings are rooted in the reef and its iron structure rises above the water. It looks octagonal from above, and like an inverted cone from the side. Most of the structure is a skeleton frame of metal columns and beams without walls, but there is an enclosed section. Red with white shutters, the enclosed part sits about a third of the way up the cone, which is hopefully above even the largest waves. This was the home of the lighthouse keepers and their assistants. Imagine lying down to sleep directly above a reef. Perhaps parrotfish slept right below. On night dives elsewhere I've seen parrotfish in protected spots between mounds of coral, each swaddled in a blankie made of mucus that they secrete each night. It's not fully understood why some parrotfish surround themselves with mucus as they sleep. Perhaps it anchors them in place or acts as an invisibility cloak, protecting the fish from predators by masking their scent.[6] The Coast Guard automated the light, so the staff moved out in 1962 and it has since been deactivated,[7] but for 137 years there were lighthouse keepers sleeping out there with the fishes.

Most people diving in shallow tropical water visiting coral reefs hope to see a place teeming with life—corals, fish, colorful invertebrates, sharks, maybe a turtle or two. But the skeletons and shells of the dead are in reefs too, even the healthiest ones. These animals don't have the hang-ups that we humans do about living alongside skeletons of the dead. Everything living in the reef is called the life assemblage. The remains of the dead are called the death assemblage. Large corals that died and remain cemented in place are part of the death assemblage, and so are broken fragments of larger colonies found in the low areas between corals.

In my search for the coral death assemblage in Florida reefs, I learned where to look. The occasional sandy areas between heads of coral in a patch reef are good spots, although they usually come with a reef shark resting on the sand. The sharks firmly stood their ground as I rooted through the first couple of inches of sediment and extracted fragments of coral skeletons, but they didn't bother me and I didn't bother them. On the shallow, high-energy reef crest, the death assemblage was often not in sand, but instead in gravel patches on either side of the row of elkhorn corals, living and dead, which were thankfully shark-free.

I had been a certified scuba diver for almost seven months when the project started, having taken the course during my junior year of college. My only prior research diving experience was assisting with a small project surveying crinoids—relatives of sea stars with long, lacy arms. I was a very green scuba diver. There's a lot they don't teach a person in an introductory scuba course when it comes to actually working underwater instead of watching marine life. I did learn how to stay neutrally buoyant, so that I could hover like a ghost above the seafloor and not touch the fragile corals on the bottom—an important skill. But I needed to know how to continually reach to the seafloor to collect the death assemblage. How was I going to hover like a ghost above the reef and also collect from the bottom? How would I carry a heavy bucket of coral skeletons in mesh bags and make sure I did not sink to the bottom?

It took me a few days to figure out how to collect without bumping into living coral. My strategy was to keep my flippered feet high above the rest of my body, my torso curved like a scorpion below. It looked silly, perhaps, but in this odd position I could kick my flippers without kicking coral. In one hand I carried my bucket and mesh bags. I extended the other hand into the reef rubble and picked up pieces of dead coral skeletons, avoiding the living reef on either side. It was like playing the game Operation®, except underwater in a coral reef, with fragments of coral skeletons standing in for the plastic bones that one retrieves with tweezers in that game.

To the snorkelers looming above me, it probably looked like I was pilfering the reef. The site was protected, yet there I was looking like the villain of a very unlikely coral heist. Instead of a sack with a dollar sign

on it, I had a mesh bag. Instead of a dark mask I wore a scuba mask. They didn't know that we had a permit to collect dead corals shed off the reef as part of a research project. Communication is limited underwater, at least with conventional scuba technology, so I had no way to explain that investigating the death assemblage in modern environments is a strategy for understanding what gets preserved and the history of the reef. Eventually, they moved on. I hope they're not still angry.

I collected 128 samples from four shallow reefs in the Florida Keys, each with enough coral fragments to add up to about the size of a bowling ball. We boxed up all the samples and sent them to Massachusetts and the lab at Smith College where I would spend months analyzing how each piece of coral changed since it died.

~

In theory, the death assemblage samples I collected could become fossils someday, but it's not easy to become a fossil. Nearly nothing fossilizes even though everything dies. Flesh decomposes after death. Bones, shells, and teeth are much more likely to be fossilized than flesh, but it's still rare. Only a small fraction of life is preserved in the fossil record—far less than 1 percent according to estimates.[8] And only a small number of those fossils are ever found.

A game called "Will you become a fossil?" in the Deep Time exhibit at the Smithsonian National Museum of Natural History illustrates the slim odds of fossilization. The game invites players who step up to the touchscreen to choose what animal they want to be from options like a trilobite, a dinosaur, or a jellyfish. When I played the game, I chose to be a trilobite, reasoning that because I have seen so many trilobite fossils in Paleozoic rocks (they lived between about 520 and 252 million years ago), their skeletons preserved with delicate detail, I was likely to become a fossil. But it was a short game for me. I was informed that I decomposed. Thus, sadly, I didn't become a fossil. Of the 3,147 people who played "Will you become a fossil?" the week I visited the museum, only 430 actually became fossils. The rest of us decomposed, were eaten, or eroded.

Fossilization potential—the chance of becoming a fossil—is not the same for all creatures. Animals with heavy, thick shells have a higher fossilization potential than those with delicate skeletons or none at all. For example, a coral's skeleton is large, much larger than the squishy bodies of its polyps, so it has a better chance of winding up in the fossil record than the thin exoskeleton of, say, a crab, which is easily broken and has numerous parts that often get disconnected after it dies. Living things that don't have hard skeletons or shells are rarely preserved as fossils. There is a good chance that the recent increase in macroalgae on many Caribbean reefs won't leave a fossil record because macroalgae don't have skeletons.

If you are trying to become a fossil, getting buried in sediment will improve your chances. All potential fossils within a death assemblage can break down over time if they aren't buried. There are exceptions to this rule—for example, if you fall into an ice sheet crevasse, a tar pit, or tree resin, you may become fossilized without getting covered in sediment—but those are not options underwater. In tropical marine environments such as growing reefs, coral skeletons and shells that get covered by sand or by the next generation of growing corals have an easier path to becoming fossilized.

Once deep enough under the sediment, the skeletons are out of the reach of animals and other organisms that might destroy them, and they are no longer broken apart and abraded by waves. After the death assemblage is deeply buried, water flows between the sediment and through the porous shells and bones, bringing the ingredients for mineral crystals, which can form within a fossil or even replace the skeletal material. Young fossils, like those found in Pleistocene reefs, are often preserved without mineral replacing the shells and skeletons. Crystals form between and within them, preserving the hard bits of dead organisms together in a tomb of rock. In much older fossils, both the space that was once home to animal tissue and the skeletal material can become completely replaced by other minerals.

If sediment accumulates gradually, the remains of dead animals that lived at different times might wind up fossilized together in the same rock. Much as a cemetery includes graves of people who died a hundred years ago and graves of people who died this year, one layer of rock can include fossils from animals that lived and died at slightly different times. For example, odds are that the shells you find on a beach are from snails and clams that didn't actually live at exactly the same time. It's likely that they didn't live in the same place either, since their shells were tossed onto the beach by waves. (I can't think of another death assemblage that people collect while relaxing on vacation and use to decorate their homes. Seashells are our favorite death assemblage.)

But corals, unlike most clams and snails, usually encrust onto the surfaces of a reef, which keeps them firmly affixed to the place where they live, even after they die. When reefs grow upward over time, younger corals grow atop the previous generation, making it possible to deduce the order in which corals lived. It's not a perfect record. Coral branches break during storms and can be moved by currents and waves far from where they lived. When dead coral skeletons are not quickly covered by more corals or sediment, they can break apart on the seafloor as they are battered by waves, chewed apart by the bioeroders, and coated in encrusting organisms, which we'll meet in a few pages.

Some rare fossil deposits do capture the remains of life with little or no alteration, like snapshots of the past. These fossil-filled layers of rock, called Lagerstätten, can form when life is buried all at once and even preserve life that doesn't have hard parts like skeletons. Lagerstätten can also form in environments deprived of oxygen, where there are few organisms that break down dead animals, plants, and other life, and in calm conditions where remains aren't broken apart. But the fossil reefs that we've been exploring aren't Lagerstätten, so not all species that lived in these ecosystems were fossilized.

When some species are fossilized and others aren't, the fossil record becomes biased. Just as the history of people with power and wealth tends to be recorded more often than the history of people with less privilege, stories from geologic history, as told in fossils, tend to favor the species with thick shells that are likely to be preserved. The more we know about the bias, the more we know which species are overrepresented and which are underrepresented in the fossil record. Our interpretations of past ecosystems are better when we know what's missing.

In fossil coral reefs, the parts that are unlikely to fossilize include the soft-bodied life that doesn't make shells and skeletons out of calcium carbonate. This includes seaweed, worms, sea fans, sea cucumbers, and microbes. But what is well preserved are corals that make sturdy skeletons, as well as the smaller calcium carbonate shells of clams and snails, spines of urchins, and flakes of *Halimeda* algae. The only evidence that can help identify fish in fossil reefs is often their tiny teeth.

Taphonomy, the study of how fossils are preserved, can help us understand whether the stories preserved in the fossil record are biased. It can also help us understand what ancient environmental conditions were like at the time when creatures died.

The field of taphonomy was named by Ivan Efremov (also spelled Yefremov in English), who was born in the Russian Empire in 1908 and died in 1972 when it was the Soviet Union.[9] He was both a paleontologist and one of the most prominent Soviet science fiction authors of his time.[10] While he was writing sci-fi stories that were often set in a distant future, he was also digging up fossils from the distant past with his wife, paleontologist Elena Dometevna Konzhukova.

Ivan Efremov described the problem of the incompleteness of the fossil record in a 1940 article, and his concern that paleontologists were drawing conclusions from fossils without considering how they were preserved and what was missing. "There is another way to the knowledge of the animal world of the past eras," he wrote, suggesting that how fossils are preserved can provide clues to what the environment was like at the time. "Taphonomical research allows us to glance into the depth of ages from another point of view."[11]

But research into how fossils form was in progress before the field had a name. In 1928, German geologist and paleontologist Rudolf Richter was studying modern environments to figure out how fossils form. He called his field of study *Aktuo-paläontologie* (actualistic paleontology or actuopaleontology),[12] a name that didn't really take off, perhaps because combining "actual" (i.e., the present) with "paleo" (i.e., the past) was a bit of an oxymoron. While the term isn't common, the concept—looking at preservation of potential fossils at the current end of geologic time and whether they are likely to make it into the fossil record—is essential for understanding how fossils are changed after death and whether the fossil record is accurate. It is another way that uniformitarianism is applied to understand fossil reefs, and it's why I collected so much of the death assemblage from Florida reefs.

～

I imagine that at the start of their voyage, the boxes of dead coral skeletons I collected from Florida reefs seemed like any other cardboard boxes in a brown UPS truck. But by the end of their trip, as the smell of marine decay strayed beyond the cardboard, the driver may have been speeding a bit in order to get those boxes delivered. When I opened the boxes in a Smith College geology lab where I planned to spend the summer documenting how the coral had been altered on the seafloor, the stench was not unlike that in the garage of my childhood house, where my sister and I stored our treasures collected from Cape Cod beaches.

I washed the ton of coral rubble in a bleach solution, replacing the decay smell with a Clorox® smell. The coral fragments, now bright white, filled the lab tables as they dried on paper. They looked like living corals that had lost their zooxanthellae in overheated water, which is known as coral bleaching, a phenomenon that has nothing to do with the cleaning product. When corals bleach, their skeletons are visible through their transparent bodies. But the death assemblage on the lab tables included no living coral. It was just the skeletons of the dead. I would be documenting evidence that these skeletons spent time on the seafloor after death, and the ways that each had been worn or broken apart—the forces that can prevent skeletons from fossilizing or obscure their identity in the fossil record.

Some of the coral fragments looked brand new, as if they had died the day before, but the surface of many fragments was worn smooth. Dead coral skeletons can be abraded by sand as waves and currents move them around the seafloor. They can be dissolved in slightly acidic water, which will become more common as more carbon dioxide infuses into the ocean from our fossil fuel emissions. If not totally dissolved or physically broken apart in waves, a coral skeleton can become a home for animals and other living things that remodel it, eventually obliterating the potential fossils. There are invertebrates that bore holes through dead coral (a process called bioerosion) and creatures that coat its surface (a process called encrustation).

Before it's buried, a death assemblage can be full of life. The more time it spends on the seafloor, the more creatures take up residence on and within the skeletons. As these creatures make themselves at home, they also destroy a potential coral fossil. Yet many of them leave their own fossils behind—evidence that they were there—which, if the coral fossil survives, can tell us even more about a fossil reef.

Many of the creatures that don't swim in a reef are adapted to encrust over surfaces, adding an outer coating to the coral death assemblage. An unusual example of what some of these creatures can do sits not far from where I collected the death assemblage in Florida reefs: an underwater statue called Christ of the Abyss. It was installed in the Florida Reef Tract in 1965, a gift from the head of the dive gear company Cressi, and marine creatures have been calling it home ever since. In this depiction, Jesus's arms are outstretched. He looks upward with a solemn expression. His arms, body, and head are covered with blotches of bright orange, yellow, green, and purple encrusting life—the marine version of graffiti.

In the lab, fuchsia spots peppered many of the coral skeletons. These spots—the skeletal remains of an encrusting foram called *Homotrema rubrum*—are common in shallow reefs. There's evidence that these forams, like the coral they attach themselves to, have symbiotic algae within them.[13] On some tropical islands, so many fragments of these forams wash up on beaches that the sand looks pink.

The forams were the most obvious because of their bright pink color, but they weren't the only species encrusting the coral death assemblage. There were also delicate lacy skeletons of bryozoans, limestone tubes left behind by worms, and the shells of clams that had cemented themselves in place. However, the most omnipresent encruster on the dead coral was coralline algae. Unlike most types of algae, coralline algae create calcium carbonate, which can fossilize. It also binds reefs together. Some pieces of coral skeleton were so completely coated with coralline algae that it was tough to identify the species of coral. As with pretzels enrobed in chocolate, only the general shape of the coral was still clear.

Other reef life, called bioeroders, are able to drill into or carve away limestone. Bioeroders can be very industrious and can obliterate a death assemblage given enough time. A study in the US Virgin Islands calculated that, annually, nearly 60 percent of the potential fossils produced in a reef are reduced to sediment by bioeroders.[14]

To look for bioerosion is to look for negative space. Bioerosion is preserved as holes in a skeleton or rock, or any area that's been carved away by a living thing. Grazing fish and urchins that take chunks from dead coral as they bite and scrape algae off its surface are bioeroders. Small animals often bioerode to make a home in limestone. Some of the most common bioeroders on reefs are boring sponges, which are boring because they bore holes in dead coral skeletons, not because they are dull. These sponges have the power to break rocks. Through chemical changes, they weaken the limestone and then physically break it apart. The evidence of bioerosion—the scattering of holes—is a trace fossil, indirect evidence of past life. I found some coral skeletons so riddled with boring sponge holes that they were more like sponges wearing dead coral suits. There was other evidence of bioerosion in the death assemblage too, such as holes left behind by worms and clams that can drill through solid limestone and live within it.

If you are a very small creature, a piece of dead coral is a wonderful habitat—that is, until you are plucked from the reef, put into a cardboard box, and sent to Massachusetts. Years later, I still feel a bit bad about that. As someone who carries spiders outside instead of stepping on them, it pains me that I caused such destruction to this tiny life on and within the coral skeletons. And yet it's only because of that destruction that we could learn what happens to the skeletons of corals after they die.

Overall, our team found that the abrasion, encrustation, and bioerosion left a signature on the coral death assemblage, and different reef zones had different signatures. The patch reef coral death assemblage had more life within and upon it than the death assemblage of the reef crest.[15] There was more abrasion in the shallow reef zones than in the deeper forereef areas, probably because there are more waves in shallow reef zones that tumble the coral fragments on the seafloor.[16] The environment left an indelible mark on potential fossils, but, as we would learn, the coral death assemblage also held other evidence.

∼

While I was darting around underwater on a dead coral treasure hunt near the Florida Keys, my teammates were calmly scrutinizing corals along a tape measure draped over the reef. Clipboard with data sheet (on special underwater paper) in one hand and a pencil in the other, they hovered a couple of feet above the reef, taking a census of what corals were alive and what corals were dead along the line. They were figuring out whether the coral species in the death assemblage were similar to the species in the life assemblage. In theory, if a death assemblage is what's likely to fossilize, and it has the same species as the life assemblage in the same reef, then there's a good chance that the fossil record is telling an accurate story of the corals in the ecosystem. But if the death assemblage has different species, then there is something unusual afoot.

They found that despite the effects of abrasion, encrustation, bioerosion, and dissolution, the coral death assemblages did show the pattern of coral species in shallow reef zones.[17] This implies that fossil reefs like those we met in the previous chapter are reliable records of what happened to corals in the past. But curiously, the death assemblages tended to have more branching corals than were found alive in reefs.[18] This could be because branching corals grow quickly, creating more branches that break during storms, so dead branches are strewn about. Or it could be because the skeletons of branching corals are easier to identify even when encrusted with coralline algae. However, there's another explanation: in some areas at least, the death assemblage could have been telling a story of Anthropocene environmental change—about the staghorn and elkhorn corals that recently died in the Florida Reef Tract and across the Caribbean.[19]

The overabundance of dead branches was most apparent in the fore-reef, at 66 feet (20 m) deep where the death assemblage included a graveyard of branching staghorn coral, likely a casualty of the 1980s die-off of

88  CHAPTER 4

*Acropora* corals. In contrast, the team found only a small amount of living staghorn coral at the site. In this case, the coral death assemblage could have been recording a shift in ecology.[20]

On modern reefs that haven't been watched closely by scientists, we often don't have a complete record of how corals have changed in recent years, but the death assemblage can hold evidence of change like a historical archive.[21] Over the past couple of decades, scientists have been using death assemblages to interpret change in modern ecosystems much as a coroner investigates a suspicious death. For example, the coral death assemblage at San Salvador's Telephone Pole Reef, which we visited in the previous chapter, is filled with the skeletons of staghorn corals that died in the mid-1980s. Even if it hadn't been monitored over time, it would be possible to deduce how the reef transformed by looking at the corals that died.[22]

Using dating techniques on the coral skeletons in a death assemblage, researchers can figure out when a reef transformed. This can help build a timeline of change in a reef. For example, by using uranium-thorium dating of corals in the death assemblage of reefs around Australia's Palm Islands, an inshore area of the central Great Barrier Reef, scientists identified that large branching *Acropora* corals in this area started to die in the 1920s, long before reef monitoring began in the 1980s.[23]

~

If all corals in a reef die at once—a mass mortality—the coral life assemblage is gone, and there is only a death assemblage, only skeletons. Without more corals to grow over the top, or sediment to bury them, these skeletons, left on the seafloor, are encrusted, bioeroded, abraded, and even lightly dissolved. The amount of degradation is a clue to which corals in fossil reefs were dead and lying on the seafloor for a while. So, a fossil reef in which all, or nearly all, the corals died at the same time would have a layer in which all the skeletons are degraded and hard to recognize.

Considering how much staghorn coral had died in the Caribbean in recent years, geoscientist Lisa Greer and her colleagues wanted to explore the fossil record to learn how staghorn coral fared in the past. They investigated a huge outcrop of fossil staghorn coral preserved in the Enriquillo Valley of the southwestern Dominican Republic. It lived 9,400 to 5,400 years ago, a timeframe that includes the Holocene thermal maximum, a relatively warm time for Earth's climate,[24] so they wondered whether the extra heat might have been detrimental to reef health as it is today. Lisa and her team documented the coral fossils and their preservation throughout a cliff of the fossil reef, using ladders to reach all levels of the outcrop. They found evidence that the staghorn coral flourished for 4,000 years without much interruption. They took a detailed look at the thickest section of staghorn coral and found that it grew continuously for at least 2,000 years. There weren't abrupt layers of poorly preserved staghorn coral covered by other coral species, so the coral hadn't died rapidly in the past as it had on reefs in the 1980s. The staghorn coral thicket had been remarkably resilient. The fossil evidence suggests that it kept growing despite the side effects of a warmer climate like possible spikes in water temperature, stronger hurricanes, and rapid sea level rise.[25]

Marine scientists Bill Precht and Rich Aronson looked for evidence of mass mortality of staghorn coral somewhat less deep in time, exploring reef subfossils that formed over the past 3,800 years below a patch reef in the calm lagoon of the Mesoamerican Barrier Reef near Channel Caye, Belize.[26] Subfossils are a sort of awkward age between the death assemblage on the seafloor and fossils in solid limestone. Precht and Aronson found evidence that the staghorn coral thicket had grown consistently for the entire 3,800 years. Older coral skeletons were buried below younger ones, all relatively well preserved. There were no layers of coral skeletons that showed the poor preservation signature of a coral mass mortality except at the very top where the staghorn coral had died, much as it did at Telephone Pole Reef in San Salvador.

Bill and Rich watched the staghorn coral die near Channel Caye. In 1987 they found that what had been, just a year before, a thriving patch reef, was in the midst of a mass mortality event. About half of the staghorn coral branches, which had been a warm orange color, had turned white. They were killed by white band disease—a somewhat mysterious illness, so named because it looks like white bands around coral branches. Along with a number of aggravating factors, white band disease would eventually kill nearly all the staghorn and elkhorn corals in the Caribbean, as we'll discover in the next chapter. Bill and Rich monitored the reef at Channel Caye each year and watched the amount of living staghorn coral drop. Seventy percent of the coral had been alive before the disease hit. Within a few years the staghorn coral at Channel Caye was entirely gone.[27]

As far as we know, white band disease can't be fossilized. In fact, we don't currently have a way of tracking coral diseases over geologic time to see whether what happened to staghorn coral has happened before. However, the degraded skeletons killed by disease are potentially preservable. The staghorn coral skeletons that died at Channel Caye sat on the seafloor long enough to became abraded and bored by bioeroders and covered by a layer of lettuce coral, which opportunistically expanded in the reef.[28] The rapid decline left a signature of poor preservation on the staghorn coral skeletons and a layer of lettuce coral.[29] But because no layers of poor preservation have been found in fossil and subfossil staghorn coral thickets, there isn't evidence that mass die-offs happened before the twentieth century.[30] What's happening in the Anthropocene is different.

In the decades since the staghorn coral at Channel Caye and elsewhere died in the 1980s, the number of problems affecting reefs, and the intensity of those problems, have increased, sending more reefs into decline, including the Florida reefs where I was collecting the death assemblage.

In the mid-1990s, there was living staghorn coral in the shallow Florida reefs that we studied, although more of it was dead than alive. Overall, these reefs had about 30 percent live coral cover at that time.[31] Ben Greenstein tells me that he chose the shallow reef sites because they were similar to the 1960s descriptions of healthy reefs, with classic zonation of corals, like the ones in Jamaica described by the Goreaus. But reefs like that were becoming rare, and so much has changed in the years since. When NOAA researchers surveyed Florida reefs in 2018, they found that some had living coral on less than 1 percent of the reef. The healthiest reefs they found had 19 percent living coral, but the majority had much less.[32]

Curious about what the reefs we researched in the 1990s were like in early 2023, I joined a flotilla of snorkelers in bright yellow vests and visited two of the same reefs, the reef crest sites. Many snorkeling tour companies still describe these reefs as epicenters of marine biodiversity, which perhaps isn't a surprise since they are trying to entice people onto their tour boats.

I jumped into the rolling waves and looked down through my mask to find the seafloor littered with the remains of the dead—piles of coral branches and overturned elkhorn coral skeletons that looked like felled trees. The reef crest still stood, but it was made almost entirely of the skeletons of dead elkhorn coral, their flat branches covered with macroalgae, sea fans, and other soft corals that whipped back and forth in the waves. The living elkhorn coral branches that I used to gingerly reach between to collect pieces of the death assemblage had now joined the death assemblage. There would be no need to play the delicate game of Operation to avoid touching coral. It was all dead. And the skeletons were probably being ground up by bioeroders and coated with encrusters.

I found one large living head of boulder star coral and was so happy to see a living coral that I kept returning to it, appreciating every inch of the bulbous hillock and its community of life—the ocher coral polyps at its surface, the Christmas tree worms poking out of it, the striped sergeant major fish buzzing around it, and a lone juvenile blue tang (which is bright yellow in its youth) lurking in a recess between lumpy parts of the coral colony. I wondered how much time that coral had left.

As the last stragglers climbed back onto the boat, I looked around at the faces of my fellow snorkelers and was struck by how many of them were smiling. I appeared to be the only one grappling with feelings of loss after our time in the decimated reef. But for those who were seeing a reef for the first time, and for those who had seen only reefs like this, what we saw was no surprise.

For decades the Florida Reef Tract has faced human impacts at an unprecedented scale.[33] Population in South Florida has grown, increasing coastal development and the amount of runoff and wastewater that can potentially make its way into the ocean and reefs. Climate change has increased water temperature in the summer and fall, and with it, the risk of coral bleaching each year. About six months after I snorkeled over Florida reefs in 2023, unprecedented heat would cause widespread coral bleaching. The heat was particularly intense in shallow, nearshore locations—the water temperature at one location was not unlike a hot tub.[34] Both pollution and climate warming make corals more likely to contract diseases. Florida's reefs were ravaged by white band disease in the 1970s and 1980s, and since 2014 they've been contending with stony coral tissue loss disease. The reefs have also been vulnerable to hurricanes, winter cold snaps, and boat damage.[35] And there's evidence that as carbon dioxide emissions dissolve into water and the ocean grows more acidic, coral limestone is dissolving away.[36] Some reefs are doing better than others, but overall, Florida's reefs are not like they used to be.

And it's not just Florida. Throughout the Caribbean, reefs have been struggling. In the next chapter we'll find out how the problems started. We'll look for evidence from recent decades and centuries that helps explain when these reefs first started to falter.

## CHAPTER 5

# WHEN REEFS FALL APART

While reefs are threatened worldwide, the Caribbean could be considered the epicenter of coral reef decline, as the region's reefs started struggling earlier than those elsewhere. This is of course terrible, but looking at what went wrong in the Caribbean over decades and even centuries provides an illuminating example of what not to do to coral reefs. Understanding what not to do can teach us what we need to do to help reefs worldwide be more resilient now and in the future.

In the last three decades of the twentieth century, researchers watched the amount of living coral on most Caribbean reefs drop by more than half, with most of that change happening from the mid-1980s through the 1990s.[1] The major reef-building coral species were hard hit. Between 80 and 98 percent of staghorn and elkhorn corals—which had dominated Caribbean reefs—died, and both species are now listed as critically endangered on the Red List of Threatened Species maintained by the International Union for Conservation of Nature (IUCN).[2] In all, 23 coral species in the Caribbean are now listed as critically endangered or endangered on the IUCN Red List, including major reef-building species.[3]

Meanwhile, the amount of macroalgae on Caribbean reefs has more than tripled.[4] Macroalgae, as mentioned earlier, are seaweeds found in oceans worldwide and include the kelp that forms forests in the cool waters off California, the sargassum that floats in the center of the

Atlantic (and is increasingly found in the Caribbean, too), and the slippery algae common in New England tide pools. While not all reefs in the region follow exactly the same pattern, and some reefs have remained healthy, the trend is clear: there has been an astonishing shift in Caribbean reefs over just a few decades, transforming the ecosystem into something with less limestone-producing coral and more of the slimy macroalgae. While Caribbean reefs got a head start, some Pacific and Indian Ocean reefs now show a similar trend.[5]

If you were taking eyewitness statements from people who saw the changes in Caribbean reefs starting in the 1970s, you would quickly identify the most obvious culprits. Your lineup would include white band disease, which attacked staghorn and elkhorn corals, a massive die-off of long-spined sea urchins in reefs, and several hurricanes from which reefs did not recover. Climate change, a dire threat today, was not causing mass bleaching on Caribbean reefs until the 1980s, and bleaching became increasingly severe in the 1990s and beyond,[6] so it may not have been on the early list of suspects (although water was already warming, which likely worsened other, more visible problems).

But the story of Caribbean reef change, as we'll see in this chapter, is one that cannot be entirely explained by eyewitnesses. That's not to say that those who witnessed the changes are untrustworthy—people did see a heartbreaking decline of reef corals and the smoking guns explaining reef ill health. But what they couldn't see were the changes in reefs that started decades to centuries before people were closely monitoring reefs. To figure out how reefs got to this point, researchers turned to records of human history and subfossils preserved below reefs to look for evidence deeper in time.

The visible problems, as we'll see, were not the only problems facing Caribbean reefs. Other reef threats remained in the shadows and were discovered in hindsight. But let's start by reviewing the timeline of the most visible changes.

According to those who witnessed the destruction, it began with disease. They found elkhorn and staghorn corals with white bands around their typically yellow and orange branches starting in the late 1970s.[7] The symptoms would start small—just a little cummerbund of exposed white skeleton where the tissues of coral polyps were completely destroyed. That cummerbund would grow wider as more polyps were killed, potentially wiping out the entire colony and then all the coral colonies of a reef. This is what happened to Telephone Pole Reef on San Salvador, Bahamas (which we visited in chapter 3), and to the reef at Channel Caye, Belize, that Bill Precht and Rich Aronson studied and watched die in the previous chapter. Like most coral diseases, white band disease is named after the effects it has on the coral rather than a specific pathogen because the pathogen is unknown (although researchers have identified bacterial suspects).[8] There's evidence that some coral diseases are caused not by specific pathogens, but instead by large groups of microbes or by bacteria that are part of a coral's microbiome, normally providing benefits rather than causing harm,[9] which we'll explore in chapter 9. White band disease still exists today, although there are fewer corals for it to infect, and some corals now appear to be less vulnerable to the disease.

Then, in 1983, people noticed that long-spined sea urchins (*Diadema antillarum*) were starting to die. The urchins had been numerous in some Caribbean locations, punctuating reefs with black starburst shapes. Urchins, which are relatives of sea stars and sand dollars, look like pincushions on the seafloor. Their skeletons are made of limestone plates with spines extending outward in all directions. The plates and spines fit together like a puzzle and are held in place by the urchin's skin. But in 1983 urchins were losing their spines. Some had light patches where their limestone plates were exposed, no longer covered by skin and spines. They were falling to pieces on the seafloor. By 1984, most of the long-spined sea urchins were dead across the Caribbean, presumably from a pathogen that was never identified,[10] although it may have been the same as, or similar to, the pathogen that surfaced in 2022, killing urchins in the Caribbean and then spreading to the Red Sea and beyond.[11]

Turn one of these urchins over and you'll find its mouth and teeth in the center of the underside, looking a bit like a bird's beak. With this beak, a long-spined sea urchin scrapes and eats macroalgae from reef surfaces. The urchins can graze on seagrass and other types of algae, depending on what's available, but their ability to eat macroalgae is their greatest superpower in a reef. After the death of the urchins, reefs struggled to keep macroalgae populations in check. The slimy carpets expanded.

The elkhorn coral, staghorn coral, and long-spined sea urchins were all taken out by disease. Caribbean marine life appears to be particularly vulnerable to disease. It's possible that species can't fight off infections, either through adaptations or microbiomes.[12] There might also be more diseases transported into the Caribbean, potentially blowing in with dust-laden wind from the desert in North Africa[13] or arriving in the ballast water of ships.[14] The urchins started dying near the Panama Canal in 1983, perhaps an indication that the agent causing disease was released in ballast water by a ship that came through the canal.[15] Alternatively, the pathogens that caused coral and urchin die-offs may have been present in the Caribbean all along but multiplied because of other environmental changes. We don't know for sure. Diseases that affect marine life, while increasingly common, remain somewhat mysterious.

Compounding the effects of diseased corals and urchins, several hurricanes in the 1980s pummeled Caribbean reefs.[16] Hurricane Allen caused disastrous damage to the reefs in Discovery Bay off Jamaica's north coast, where the pattern of zonation was first documented by Thomas and Nora Goreau in the 1950s and 1960s.[17] The hurricane passed north of Jamaica as a Category 4 storm in August 1980. By this time a new generation of scientists were studying the reefs at the research station the Goreaus had founded. They donned scuba gear and documented the hurricane damage once the storm had passed, finding that the reef's elkhorn coral had been reduced to rubble. Sand carried by the waves had blasted against the corals. Delicate branching staghorn corals were broken and overturned. Many sturdy, massive corals survived with only minor damage.[18] "A few of the colorful fish remained, but they swam at you with shredded fins and missing scales, looking like disaster refugees," wrote researcher-turned-filmmaker Randy Olson about what he saw while surveying reefs after the storm.[19]

Most of the elkhorn and staghorn coral fragments that survived the hurricane eventually died in the months after the storm.[20] Some may have been too battered to recover. Others probably died from white band disease.[21] Rotund massive corals and encrusting corals (which are not affected by white band disease) were more likely to heal and survive unless the damage was extensive. The team that surveyed the damage projected that the reef would eventually recover,[22] which makes sense given that hurricanes are natural and corals have adapted to survive in this tropical environment, growing new coral colonies from broken fragments after storms.

But the reefs at Discovery Bay haven't bounced back.

With the cascade of problems, many Caribbean reefs, including those at Discovery Bay, entered a new phase, one dominated by macroalgae instead of coral. Corals dying of white band disease opened space for macroalgae to settle down and spread out. Without the grazing urchins, macroalgae were able to grow unchecked. On some reefs, weedy or opportunistic coral species moved in when staghorn and elkhorn corals died. Weedy coral species—like lettuce coral and finger coral—don't form the nooks and crannies that fish and invertebrates rely on in a reef, and they don't have the same role in the ecosystem as corals that build the reef framework.[23]

The shift from Caribbean reefs dominated by framework-building corals to reefs dominated by macroalgae and weedy coral species is called a phase shift,[24] a change in an ecosystem in response to a change in environmental conditions.[25] In general, ecosystems that are prone to phase shifts have lower resilience, less ability to bounce back.[26] There is no known evidence of this type of reef phase shift in the Caribbean fossil record, implying that past Caribbean reefs were more resilient. However, it's hard to find evidence of resilience in the fossil record because resilience is the ability to bounce back from a perturbation, and in general it's hard to find evidence of perturbations, except those of such a large scale that they shut down reef building, as we'll see in the next chapter. What is very clear in the fossil record is that in the past, Caribbean coral reefs had an extraordinary ability to persist, rebuilding again and again with similar corals and reef zones over geologic time.[27] Today's Caribbean reefs don't appear to be like those in the past.

Many Caribbean reefs had entered this new phase by the late 1980s. Some were a bit later, shifting phase in the 1990s. Coral cover decreased from an average of 35 percent, or perhaps as much as 50 percent, down to 16 percent between 1970 and 2012, with most of that change happening between 1984 and 1998.[28] Macroalgae cover on Caribbean reefs increased from 7 percent to about 24 percent between 1984 and 1998 and has held steady since.[29]

Those averages mask the patchiness found throughout the region. Some reefs have remained much healthier than the average, with more living coral and less macroalgae. And unfortunately, other reefs are less healthy than the average, with little live coral—as in Discovery Bay, Jamaica, where coral cover is less than 10 percent and macroalgae enrobe more than 60 percent of reef surfaces.[30]

Some reefs have been able to bounce back from the phase shift over time. For example, on Dairy Bull Reef, just over a mile (2 km) from Discovery Bay, corals declined and then mostly recovered by 2005.[31] Even staghorn coral made a comeback, filling the spaces between massive corals with tangles of branches.[32] The return of those spiky urchins is credited with the recovery of the corals. In areas where long-spined sea urchins have returned, there is more coral and less algae.[33]

Although it appears that macroalgae invaded the Caribbean, algae aren't evil. They are a normal part of coral reefs. The problem with the phase shift is the proportion of macroalgae. There is evidence that macroalgae make life harder for corals, for example by harboring pathogens that cause coral disease, and by preventing coral larvae from settling down and starting new colonies. Additionally, reefs with more macroalgae and less coral tend to have less topography, which is built mostly by coral skeletons and coralline algae. With less topography, there are fewer places for fish to shelter, and reefs are less able to buffer a coast during storms. From an economic perspective, reefs filled with macroalgae tend to attract fewer tourists, who prefer to snorkel and dive on healthy reefs. In the future, reefs dominated by macroalgae or weedy coral species like lettuce coral may not grow upward fast enough to keep up with sea level rise. Coral reefs will eventually die if they are unable to grow upward enough to stay within sun-drenched depths (as we'll discover in the next chapter).

While Caribbean reefs are particularly prone to booms in macroalgae cover,[34] reefs in other parts of the world are not immune to phase shifts.

Parts of Australia's Great Barrier Reef near the coast may be undergoing a phase shift toward less coral and more macroalgae as well.[35] Between 1985 and 2012 about half the coral cover on the Great Barrier Reef was lost.[36] Nearshore reefs have been particularly affected, while more remote reefs appear to be able to avoid phase shifts, perhaps because they aren't contending with as much environmental change as the nearshore reefs.

The Great Barrier Reef did not experience the same stresses as Caribbean reefs decades ago, except for hurricanes (which are called tropical cyclones in the Pacific).[37] Coral diseases, while present, are not decimating reefs in the Indo-Pacific region as they have in the Caribbean. But acute threats to Great Barrier Reef corals include widespread bleaching and outbreaks of the crown-of-thorns starfish. It's estimated that crown-of-thorns starfish have caused nearly half of the coral loss in recent years.[38]

A crown-of-thorns starfish can grow to be larger than a dinner plate and have up to 21 arms, each lined with spines. It might be red, pink, orange, green, blue, or bright purple and would look as festive as a Muppet if it didn't cause such devastation—these sea stars specialize in eating corals. They trundle over coral colonies like a reef Roomba®, gobbling up polyps. When starfish numbers are low, corals grow faster than they are consumed. But with too many starfish, a reef is left with dead coral skeletons that can become covered in macroalgae. Starfish outbreaks were first spotted in the 1960s. The mechanism is still debated, but it appears that outbreaks happen in areas where runoff from agriculture and cities adds excess nutrients to seawater, which is more common in nearshore waters than offshore.[39]

While there's evidence that the Great Barrier Reef is heading into a phase shift, with changes in ecology caused by these stresses, there is also evidence that at least parts of it are able to bounce back and avoid a perpetually algae-dominant state. In August 2022, the Australian Institute of Marine Science released its annual report on reef condition, which contained some good news: the percentage of coral cover in the northern and central parts of the reef had increased to its highest level in 36 years.[40] There were less acute disturbances and the coral cover was able to expand. However, the increase in coral consisted mainly of species vulnerable to heat stress, starfish outbreaks, and tropical cyclones. "Future disturbance can reverse the observed recovery in a short amount of time," noted the report's authors.[41] In other words, don't get too comfortable. Resilience has its limits.

⁓

The observed changes in the Caribbean build a story like this: we invented scuba and saw healthy reefs, but then corals and urchins started to die from disease, hurricanes hit, and many reefs were left with more macroalgae and less coral. This fits the observations but assumes that the first scuba divers actually saw healthy reefs. Perhaps they didn't. Evidence preserved in subfossils and historical records, as we'll see, suggests that Caribbean reefs were already coping with change, their resilience compromised long before we met them directly through the glass of scuba masks.

When you see a new place, a coral reef for example, it forms an impression. And that impression serves as your baseline for what reefs are like. It defines what is normal. Your baseline doesn't take into account what happened in a coral reef before you saw it. Any future changes would be compared to how reefs appeared when you first got to know them, which is why I was so sad to revisit Florida reefs in 2023 and see them looking terribly degraded compared with the reefs I had met nearly three decades before while collecting the death assemblage. Those who arrive to see reefs in the future would likely form their understanding of what's normal at that point, which may be algae-covered reefs with weedy corals.

This concept is known as shifting baseline syndrome, which sounds like a medical condition but actually describes what happens if environmental knowledge is not passed from generation to generation. If people experience only unhealthy ecosystems, they don't know what's missing so they won't know what could be restored.

Marine biologist Daniel Pauly first shared the idea of shifting baseline syndrome, relating it to fisheries.[42] He proposed that if we assume the ocean was natural when we first got to know it, we are ignoring how our history has changed it. We are not just spectators of fish. We affect them. Because of overfishing, many fisheries are depleted. Some were depleted before we started keeping track of fish populations, and our conservation goals need to take this into account.[43]

With coral reefs, shifting baseline syndrome can be applied a bit differently because reefs are an ecosystem instead of populations of fish, and because the window was opened to fully observe reefs in detail only once modern scuba diving equipment was available,[44] with a few exceptions. One of these exceptions was a team from the American Museum of Natural History that, in 1924, got to look at a reef from within a tube that dipped down to the shallow seafloor near Andros Island in the Bahamas.[45] It looked somewhat like a human-sized hamster Habitrail, according to an illustration of the technology.[46] Museum staff made observations and painted watercolors of the reef from within the tube (I really wish I could have been there with them). Between 1923 and 1933 they also dove to the seafloor wearing massive, 65-pound steel helmets tethered to a boat by hoses supplying the diver with air. Wearing these massive helmets, they collected 40 tons of coral from the Andros reef, which was shipped to New York City for a coral reef diorama at the museum (we now know that this scale of collecting is terrible for reef health).[47] Once scuba technology was available, scientists including Thomas F. Goreau started diving in coral reefs, describing them as never before, and an entire branch of science collectively saw a new place for the first time, cementing the idea of what a normal coral reef was like.

Studying reefs today almost certainly means studying an ecosystem marked by Anthropocene change, but decades ago it was thought that reefs with classic zonation were pristine. Marine ecologist Jeremy Jackson disagreed. He postulated that humans had been changing Caribbean coral reefs long before we saw them. In a 1996 presentation titled "Reefs since Columbus" at an international coral reef conference, he made the case that Caribbean reefs had been coping with human impacts for hundreds of years and that during the wave of European conquest and colonization that followed Columbus's arrival in the Caribbean, life in the ocean was changing just as life on land was changed by the colonizers.[48] This idea ushered in a new way of thinking about Caribbean reefs.

Pleistocene reefs lived without much, if any, human influence. But in more recent history, humans and reefs coexisted: reefs were just under the shimmering sea surface as we lived on land nearby. Reefs were just offshore when Indigenous populations were the main human inhabitants of the Caribbean—for example, Lucayan communities were living near Telephone Pole Reef on San Salvador and throughout the Bahamas, and Maya communities were living near the Mesoamerican Barrier Reef. Reefs were still just offshore when scores of Indigenous people were killed or enslaved after Columbus reached the Caribbean in 1492. Reefs lived alongside colonists as forests were toppled to make way for plantations, and as populations on Caribbean islands swelled with European colonists and African and Indigenous people enslaved on those plantations, a system that was not fully abolished until the nineteenth century. Reefs were just offshore as large-scale agriculture, cruise ships, and global shipping became common in the twentieth century, and as tourists arrived on tropical vacations. Because corals can be long-lived, some coral colonies alive today may have survived all of this even as many neighboring corals died.

If we didn't get to know reefs when they were pristine, then we didn't see firsthand when they started to change. Eyewitness testimony doesn't tell the full story of what happened to Caribbean reefs. But there are clues that fill the gaps in our understanding between what's found in fossil reefs and modern reefs, connecting geology and history, and these clues can help us figure out when and why Caribbean reefs started to decline.

Historical ecology, a relatively new field of research, looks at how ecosystems fared in the recent past, before we were monitoring them yet after our actions may have had an impact. It focuses on the interactions between humans and ecosystems over time and how changes in human societies affect the natural world, often leaning on a combination of historical records and not-quite-yet fossils. The research on death assemblages that we learned about in the previous chapter can help us understand change over recent decades. And research on sub-fossils, the remains of past reef life that are buried below modern reefs, can be useful for looking at evidence of change in reefs over many centuries. If a living ecosystem sits at the shallowest part of deep time, historical ecology is like the deep end of the kiddie pool—further back than the present day yet much more recent than the Pleistocene.

According to historical ecology research, what looked like the culprits causing reef decline around the 1980s actually may have been just the last straws. The story of whodunit is more complex, involving multiple threats to reef health over hundreds of years, including both long-term chronic problems and more brief, acute disturbances. As we'll see, what happened in our human world above sea level affected the world of coral reefs below.

---

One clue to what reefs were like before scientists began studying them closely is the word "corals" written in elegant script on old nautical charts. Not all chart makers added ecological observations to their charts, but George Gauld did. An engineer and artist, he charted the shallow waters around the Florida Keys in the late 1700s for the British government, which was keen to have charts of its newly claimed territory.

Nautical charts, maps that help sailors navigate, include features like the shape of the coastline and the location of lighthouses. They indicate hazardous sandbars, rocks, and—in the tropics—coral reefs. Water depth is an essential component on nautical charts so that ships avoid running aground in shallow water. The water depth measurements on the first nautical charts of the Florida Keys show how frequently George Gauld took soundings—there are so many that they form nearly dotted lines along his ship's path as it zigzagged around the shallow water.

His method for taking depth soundings in the eighteenth century would have been much like Captain FitzRoy's method for taking soundings around the Cocos Keeling Islands in the nineteenth century—a sounding line with a lead weight containing sticky tallow. The charts included notations about what was on the seafloor based on what the tallow picked up, and the word "corals" was added to the chart where the tallow picked up the impressions of coral instead of sediment. In addition, George Gauld noted what he could see on the seafloor through shallow, clear water. He added "brown coral banks" to charts where he could peer over the side of the boat and see brown coral below, which may have been elkhorn coral.

Because of Gauld's charts, we have a record of where corals were located in the waters around the Florida Keys about 200 years before Florida reefs were first monitored by research divers and instruments. There's a lot he wouldn't have known, such as whether the corals were living or dead, but the charts give a glimpse of what the reefs were like more than two centuries ago and tell us that many of the area's corals were lost long before we ever got a good look at them.[49]

Loren McClenachan, a historian and marine scientist, found these early charts of the Florida Keys when she was mining archives for Caribbean ecological information. Most people go into an archive with a specific time period and geographic location to research, but Loren wanted to know about a large geographic area (the Caribbean) over a long time (hundreds of years).

"It was like looking for a needle in a haystack," she recalls when I ask about her research process. She was searching for mentions of sea turtles, monk seals, manatees, and other marine life within giant tomes and stacks of papers, records from colonies and expeditions to the Americas written in script with old-fashioned language. She encountered the nautical charts at the UK Hydrographic Office Archive.

Early nautical charts have much more natural history in them than modern charts, according to Loren, and George Gauld's charts, especially, included environmental and ecological information that went beyond what a mariner might need to know.

"I think he just had an eye for the natural history because his charts are just beautiful," she tells me with admiration for the chart maker. In addition to corals, he included other ecological notes too, such as where he saw the crawl marks of nesting turtles in beach sand.

Loren and her team compared the eighteenth-century charts with modern maps in order to understand how reefs have changed in the area around the Florida Keys over 240 years. The modern maps were made with satellite data collected by instruments that can parse what's on the seafloor at shallow depths. The team found that corals indicated on the old charts were often in places where there are no corals today, places where we've never known corals to live.

Overall, 52 percent of the corals on the old charts no longer exist according to current maps.[50] That loss was not equally spread across the area. Reefs offshore are largely intact, although live coral cover has dropped, which is not something the nautical charts could document. But reefs near the coasts of the Keys and in Florida Bay are mostly gone. About 69 percent of corals have been lost from nearshore reefs. About 88 percent of corals have been lost from Florida Bay. The researchers call the missing and forgotten corals "ghost reefs."[51]

～

While Loren McClenachan's historical ecology research looks into archives to discover what happened to reef ecosystems in the past (along with old nautical charts, she has found evidence of past marine life in the records kept by pirates, European colonists, and tourists, all of whom unintentionally documented aspects of ecosystems), marine ecologist Katie Cramer looks below the surface of reefs for evidence of how and when reefs changed over history. She's found that reef subfossils—skeletons and other remnants of reef life that are between fossils and the death assemblage in age—hold numerous clues to how reefs and humans coexisted in times ranging from decades ago to several thousand years in the past.

Growing up in the American West during a time of rapid urbanization and development, Katie saw changes everywhere in the landscape and noticed that it can be easy to forget what has changed. Both Katie and Loren began their historical ecology research while working on their doctorate degrees with Jeremy Jackson, who first applied the concept of shifting baselines to coral reefs and argued that the first reefs we saw may not have been as healthy as they once were.

To get some perspective about changes in Caribbean coral reefs before we were monitoring them, Katie and several collaborators compiled previous research on corals from the Pleistocene to the present, creating a timeline of what's known about Caribbean reef corals.[52] The oldest coral data came from fossil reefs at the fringes of islands and coasts like the Cockburn Town fossil reef on San Salvador. Somewhat younger data came from historical records and subfossils of past reefs extracted from below living reefs. Modern data came from underwater surveys collected in recent decades since the advent of research diving.[53]

According to their census of Caribbean coral species through time, Pleistocene coral communities were for the most part similar to the coral communities that Thomas F. Goreau found when he dove into Discovery Bay, Jamaica, in the 1950s and 1960s. However, the corals in reefs started to change at least a couple of decades before we were likely to notice. Elkhorn coral started to decline in or before the 1950s, at least two decades before it would start to be decimated by disease. Staghorn coral started to decline about a decade after elkhorn coral.[54] It's challenging to know exactly how early the changes began because data are limited, but what's clear is that white band disease and the mysterious urchin illness were not the only agents changing Caribbean reefs. Katie thinks a drop in water quality started the midcentury decline in coral health.[55]

Much as the term "air quality" describes the level of pollution in air, "water quality" describes the level of pollution in water. Water quality in the shallow coastal ocean depends on what is happening on land because water that falls on the land runs off into the ocean if it isn't absorbed into soil and rock.

Even clean and clear freshwater runoff from land has an impact on coral reefs. It makes seawater brackish and carries sediment, neither of which is an ideal environment for coral. Sediment clouds seawater, allowing less sunlight to get to corals for photosynthesis. Sediment is also a problem for corals feeding on plankton because they need to expend extra energy to clear sand grains away from their tiny mouths. As a result, reefs rarely form where sediment-laden rivers meet a coast. Charles Darwin noticed this pattern as he documented the breaks in barrier reefs where rivers entered the ocean around remote Pacific islands during his voyage on the *Beagle*.[56]

In places where land has been reworked for agriculture and developed into cities and towns, changes in runoff can transform reefs. The amount of runoff increases when buildings and paved surfaces cause land to absorb less water. And the runoff itself can contain nutrients and chemicals that are hazardous for reef life. It's likely that the ghost reefs on the old Florida nautical charts were destroyed in part by runoff from land,[57] since they were mostly near shore and in Florida Bay, areas that are most exposed to runoff. "When you think about the history of change in that place, it makes sense that the impacts would be inshore," says Loren, noting that the reef declined around the same time that the Keys were becoming populated and connected to the mainland by a railroad and, later, a highway. This was also around the time that much of the Everglades was converted from wetlands to agriculture and urban development, which sent more pollutants and less water into Florida Bay.

Katie found herself in murky water while working on her dissertation off the coast of Bocas del Toro, Panama. Much of the land of Bocas del Toro province is used for large-scale banana production, the most chemically treated of all major food crops.[58] Towns and hotels that support a thriving tourist industry also pepper the coast. With tropical forests transformed into agriculture and towns, water flowing from the land to the ocean changed.[59] "It's pretty clear that the water is not the way it used to be," Katie says. In that murky water, she got curious about digging into the subfossils below the seafloor to learn how development on land affected reefs in the past.

"It really seems that something happened around the mid-twentieth century," Katie tells me when I ask when poor water quality started to cause elkhorn and staghorn corals to decline. "Right around the 1950s, synthetic fertilizers were first being deployed, the green revolution took off, and mass agricultural practices kicked in."

Across the Caribbean region, populations and tourism were growing during the twentieth century. More people meant more deforestation, development, agriculture, sewage from urban areas, and oil and other toxins, as well as more widespread use of chemical fertilizers.[60]

Runoff that carries fertilizers and urban pollution like sewage adds nutrients to seawater, which in reefs is linked to macroalgae growth and outbreaks of coral disease.[61] Runoff also makes corals more vulnerable to bleaching.[62] Higher levels of nutrients in the water allow more phytoplankton to grow. While some phytoplankton on a reef can be beneficial, too much can disrupt the ecosystem.

Runoff lowered water quality and changed reefs earlier in some Caribbean locations, particularly in places where European colonists were clearing land for plantations and towns. For example, there's evidence that elkhorn coral, which used to thrive around the island of Barbados and is found in its fossil reefs, died after sugar plantations replaced native vegetation in the mid-seventeenth century.[63] Similarly, although on the opposite side of the planet, branching corals were lost from Moreton Bay, Australia, in the mid-nineteenth century, around the same time the city of Brisbane was settled by colonists along the coast of the bay.[64] It's likely that runoff from the growing city flowed into the bay, and while some corals were fit for the challenge of living in less pristine water, the branching corals were not.[65] Near the Palm Islands in the central Great Barrier Reef, the large branching *Acropora* corals, which are now found only in the death assemblage, started to die a century ago, around the same time that colonizers cleared land for agriculture and cities, which could have decreased water quality.[66]

There was another change in Caribbean reefs, however, which likely made the ecosystem more vulnerable to phase shifts, and it began long before corals declined: overfishing. Missing from Caribbean reefs today are many of the sharks, fish, and turtles that used to live there, which has an impact on reef resilience because these animals play important roles in the reef ecosystem and, as we'll discover, can even affect the amount of coral living on reefs.[67]

Overfishing occurs when there is too much fishing—when fish are caught at a higher rate than they reproduce—which depletes their numbers. When the largest fish are sought by fishers, the population of an overfished species shifts to smaller, younger fish over time. Yet because fishers are generally proud of their largest catches, there's a unique record of overfishing in reefs. It is documented in historical photos of fishers showing off their catch in Key West, Florida.

From the 1950s to the 1980s, a photographer in Key West went down to the docks each day when boats were returning from fishing trips and photographed delighted people showing off the largest fish they had caught.[68] He would send the photos to the people pictured, who likely used them to support the stories they told about catching the big one.

Case in point: In one black-and-white photo taken in the 1950s, a family of three shows off the catch after a day of fishing in reefs near Key West. There's a man, a woman, and a girl in a frilly dress, perhaps six years old, who is holding a squirming puppy. Both the girl and the puppy look as if they could have been swallowed whole by any of the enormous groupers hanging open mouthed from a rack behind them. At least one grouper is longer from jaw to tail fin than the man is tall.

When the Key West photographer died, he donated decades of fish photos to the Monroe County Public Library's Florida Collection, which is where Loren McClenachan found them, toward the end of her search of history archives for ecological information. She had been rooting through the Florida Collection in a back room of the library for three weeks when the lone staff person in the archive handed her the box of fish photos.

"I don't know if these might be of interest," he said, according to Loren.

"Absolutely! Like, this is the best thing I've seen in two years!" she exclaimed. Loren knew these were not just photos. They were data, a record of reef fish.

The caught fish in the photos were all from the same charter company's trips, which visited similar shallow reef areas near Key West during each trip, so the fish caught reflected the largest fish living in the area's reefs. The fish were hung from the same rack in all the photos, so there was a consistent item for scale. Loren used the photos as a record of how fish in the area changed over time. The series of photos is a time-lapse of shrinking fish.

The goliath groupers were probably at their maximum size in the 1950s photos. Through the 1960s and 1970s, people's fashions changed

Captain Tony and Mae Tarracino with their daughter Coral, a puppy, and many goliath groupers caught off Key West, Florida, on September 9, 1958. Florida Keys History Center, Monroe County Public Library, photo by Wil-Art Studio, CC-BY-2.0.

and the fish in the photos got slightly smaller. The 1980s photos no longer featured groupers. Only snappers and other, smaller fish were on display. A 1980s photo shows a grinning man proudly holding a fish that is no longer than his forearm. The child next to him is a giant compared with all the fish displayed on the rack. The most recent photos, which Loren took at the same location in 2007, show fish that are on average just over a foot long.[69]

Loren identified the species in the photos, measured them, and used their lengths to calculate the approximate weight of the fish. She found that fish size dropped from an average of about 44 pounds in the 1950s to about five pounds in 2007. The largest fish hanging on the rack in the 1950s, goliath groupers, were hundreds of pounds. Reef shark populations had already been reduced by commercial fishing by the 1950s, and large groupers had been fished since the 1880s.[70] According to Loren, while the groupers were still large in the 1950s, their numbers were depleted so they were probably no longer found in as many areas as they once were.

Groupers and sharks sit at the top of the food chain in coral reefs. Without these large predators, the whole ecosystem changes. A study comparing coral reefs at Pacific islands with and without healthy shark populations found that without sharks, the abundance of prey changes, as one might expect. But there are also changes to the way energy flows through the ecosystem and how nutrients are cycled. The researchers concluded that when these predators and other fish are protected and pollution levels are low, reefs are more resilient.[71]

Reefs also need healthy populations of herbivores, including large herbivores like green turtles, which eat macroalgae and seagrass. Green turtles move between coral reefs and seagrass beds in the Caribbean and influence both environments. Because they are large, their impact on ecosystems is also large: one adult green sea turtle eats about the same amount as 500 long-spined sea urchins.[72] But the turtles were overhunted in much of the Caribbean by about 1800,[73] when islands, captured and colonized by Europeans, needed food for their growing populations. Today there are only about 300,000 green turtles left in the Caribbean. There were an estimated 91 million green turtles when Christopher Columbus arrived in 1492.[74]

Hunting a sea turtle required little skill or heroism. The female turtles came up onto the beach to lay their eggs and were taken by colonists. They move awkwardly when out of water, inching their bodies, at least 200–300 pounds, forward across the sand by rowing their flippers, determined and driven to lay their eggs. According to a 1666 report, 50 nesting females could be taken from Grand Cayman in less than three hours.[75] Historical records indicate that 13,000 Grand Cayman turtles were killed and transported to Jamaica for food each year between 1688 and 1730.[76]

The largest nesting population of green turtles in the Caribbean is currently at Tortuguero, Costa Rica, where about 30,000 nest, on average, each year. The population supported by this nesting site is an estimated 130,500 adult turtles, including males and nonnesting females.[77] In comparison, the Grand Cayman nesting site in the 1600s supported a green turtle population of an estimated 6.5 million adults—about 50 times more.[78]

Sea turtles aren't the only swimming herbivores in coral reefs. Grazing fish also help keep reef macroalgae populations in check. And parrotfish, the same reef inhabitants that crunch limestone and create sand-filled poop on limestone platforms, happen to be essential reef grazers. With the crunching sound that carries through a reef, they are both grinding rock into sediment and eating lots of macroalgae off the reef. There are other grazing fish on Caribbean reefs, but parrotfish are particularly efficient at curbing algae. When awake, they spend nearly all their time eating. In addition to the parrotfish species that crunch rock, some parrotfish just bite the tops off the algae, and some scrape algae entirely off limestone. Parrotfish also occasionally eat coral off the reef, but because they remove so much algae, they do more good than harm.[79]

There's evidence that when parrotfish numbers are high, reefs do a better job of bouncing back after a hurricane or other disturbances. In reefs of the southern Caribbean island of Bonaire, where the parrotfish population is relatively healthy, a team of researchers surveyed corals, macroalgae, and fish over 15 years at sites with remarkably high coral cover and low macroalgae until a 2008 hurricane and a 2010 coral bleaching event.[80] The dual disturbances caused coral cover to drop by more than 20 percent, and macroalgae cover to triple.[81] There were about half as many juvenile corals on these reefs after the hurricane and bleaching, yet the reefs were able to recover their coral cover and juvenile corals between 2010 and 2017. The researchers credit parrotfish for this resilience.[82] As parrotfish eat the macroalgae off the reef, they create more space for juvenile corals to settle down and for existing corals to expand. In the seven years after disturbances changed the reefs, grazing decreased the amount of macroalgae, opening up more space for corals.

Parrotfish are still common in the Caribbean—they are one of the most frequently counted fish in reef surveys; however, in many areas, they are overfished.[83] Overfishing affects nearly all areas of the Caribbean that do not have protections in place. The problem of overfishing isn't

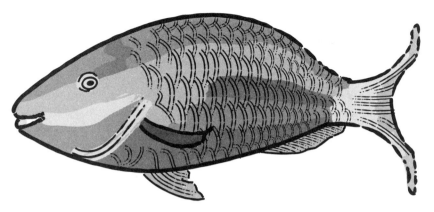

Adult stoplight parrotfish (*Sparisoma viride*).

specific to reef fish, or to the Caribbean. Worldwide, a third of fisheries are overfished.[84] Although, because of the shifting baseline syndrome, there may be more depleted fish than we recognize.[85]

To figure out when parrotfish numbers started to decline in the Caribbean, Katie Cramer's team went looking for fish teeth in the subfossils that accumulated in reefs. Assuming that the sediment and the fish teeth within it are well mixed, the teeth should reflect the number of fish on the reef. Working with the Smithsonian Tropical Research Institute, they extracted cores, four-inch-wide cylinders of subfossils, by hammering a device that looks like a giant metal drinking straw into the seafloor near reefs in Panama and Belize and pulling the cores up with a winch. Their subfossil cores from Panama reefs extend back 3,000 years. The cores from Belize reefs extend back about 1,300 years.[86]

To extract fish teeth from the reef cores, Katie turned to a method that her collaborator Dick Norris had used to locate fish teeth in cores of deep-sea sediments. The team sliced half of each core into five-centimeter increments, which became the samples. After they separated out large pieces like coral fragments, urchin spines, and clam and snail shells, each sample was left with only small grains. With a bath in mild acid (about the strength of vinegar), the only items remaining were those that didn't dissolve, including fish teeth. A dye added to the sample turned the fish teeth pink, making them easy to find.

Inside its beaklike mouth, a parrotfish has about a thousand teeth, and it loses and replaces teeth throughout its life. If there were a parrotfish tooth fairy, she would be extremely busy. Even though the cores were from only a four-inch area of each reef, there were an average of 74 fish teeth in each five-centimeter sample. Nearly half of the teeth were from parrotfish. Parrotfish teeth are more substantial than the teeth of reef fish like gobies, but they are still quite small. A typical parrotfish tooth can be up to a quarter of a millimeter across, or slightly smaller than a grain of salt. It took a lot of eyes looking through microscopes to find and identify thousands of years of fish teeth from the cores, according to Katie.

There were fewer parrotfish teeth in the recent samples, indicating that parrotfish were more common in the past. The researchers believe that overfishing was the primary cause of parrotfish declines. Most of the decline in parrotfish appears to have happened after Columbus arrived in the Caribbean, when European colonists and the slaves they brought to the area started fishing more intensively. In the 1800s, the population of Garifuna people grew in Belize, which may have also contributed to overfishing. But they found two Panama locations where parrotfish numbers started to fall one to two millennia ago.[87] Nearby Indigenous communities were small at that time, so it's unlikely that the drop in parrotfish numbers was the result of human impacts, and other environmental changes likely played a role.[88]

Interestingly, the team looking at parrotfish teeth through time found that the number of parrotfish had an effect on reef coral growth. On all the reefs studied, when the number of parrotfish was high, corals grew fast, implying that reefs need parrotfish to thrive. According to Katie and her collaborators, it's likely that when parrotfish control the macroalgae on a reef, corals are able to dominate, growing faster. In Panama and Belize, healthy parrotfish populations in the past allowed coral reefs to grow limestone relatively rapidly over hundreds of years.[89] This suggests that parrotfish are to reefs what pollinating insects are to gardens—providing an important service while just trying to find their dinner.

However, the same may not be true today. Studies looking at modern reefs where parrotfish and other grazers are protected, such as in marine protected areas, have found that today's corals are not necessarily healthy when parrotfish numbers are high. For example, in Florida, corals have faced steep declines, yet parrotfish are large and healthy because there are zones where no fishing is allowed and because parrotfish are not a target for anglers or commercial fisheries in the area.[90] With so many other disruptions on Florida reefs today, including poor water quality and devastating heat stress caused by climate change, it's likely that healthy grazing fish aren't enough to help the corals.

But overall, a reef needs grazers to keep macroalgae in check. Caribbean reefs started out with fewer species of grazers than reefs in other areas of the world (just as there are fewer coral species in the Caribbean than elsewhere, there are fewer reef grazer species too[91]), so losses from overfishing had an outsized impact. Without as many green turtles, grazing in reefs was left to fish and urchins. When overfishing reduced the numbers of grazing fish, urchins became the primary agents clearing macroalgae from reef surfaces. When the urchins died in the 1980s, right when macroalgae were growing into space freed up as corals died, macroalgae boomed and the impact was catastrophic. But the urchins were just the last straw. The reefs had been short on grazers for centuries.

---

Given the confluence of problems that have disrupted Caribbean reefs over decades to centuries—from overfishing, to the influx of nutrients, to disease and heat—is there any way they could bounce back?

Many reefs do not appear to be able to recover. But to a degree, coral reefs should have the ability to bounce back from disruptions. Disturbances could even be beneficial. At the scale of a landscape (or seascape, in the case of coral reefs), disturbances create patchiness, which increases biodiversity. Disturbances can reduce the ability of some species to dominate and exclude others.[92]

Ecologist Joseph Connell concluded that disturbance could be beneficial to reefs. In the 1960s and 1970s, he found that high diversity on coral reefs at the southern end of the Great Barrier Reef was caused by periodic disturbance. Areas of the reef that were exposed to tropical cyclones and damaged by storms were recolonized by many species of coral, so they had higher diversity than areas of the reef that were not disturbed by storms. In undisturbed areas, he found that corals were able to outcompete their neighbors—for instance, by growing on top of a neighboring coral or growing quickly enough to shade their neighbors.[93]

However, it's thought that there is an ideal amount of disturbance that keeps diversity high in an ecosystem. Too few or too many disturbances and diversity will drop.[94] If a reef is repeatedly battered by disturbances like disease, hurricanes, and extreme heat events, it might not be able to recover. There might not be time for corals to recolonize and grow before the next disturbance hits. If disturbances are infrequent and short term, there's a better chance that reefs can bounce back.

Evidence of short-term reef disturbance and recovery after disturbance is rare in the fossil record, but there is one interesting example in the fossil reefs of Papua New Guinea that helps provide some perspective on how changes in reefs over recent decades and centuries compare with those in the more distant past. John Pandolfi (whom we met earlier in this book) and several collaborators found clues in a fossil reef to what can happen when disturbance kills reef corals en masse. As you may recall, the fossil reefs on Papua New Guinea's Huon Peninsula are preserved in nine stair-step terraces, formed as land uplifted. John and his team took a detailed look at the stairstep that formed in the mid-Holocene (11,000 to 3,700 years ago) and found that the reef in this location died four times over thousands of years. At least twice the reef was smothered by ash from nearby volcanoes.[95] The most widespread of these volcanic ash events, which occurred over 9,000 years ago, is preserved as a broad expanse of fossil corals draped with a thin layer of volcanic ash like a coral reef version of Pompeii.

The fossil skeletons of corals that died in these disturbances have a distinct signature—some are covered with a thick layer of encrusters and others are broken fragments, taphonomic evidence that dead coral skeletons were lingering on the seafloor for some time,[96] as we explored in the previous chapter. But above each layer of killed corals is another fossil reef. Less than a century later, corals were able to colonize and grow on top of the skeletons of the dead. These fossils are evidence of reefs that were killed, but they are also evidence of resilience, showing that coral was able to return.

The Papua New Guinea fossils show that at least some reefs had to contend with catastrophic disturbances long before the Anthropocene, but they also emphasize what's different now. John and his team found that disturbances affected these mid-Holocene reefs much less often than disturbances affect reefs today, only about once every 1,500 years.[97] Today there are a multitude of different disturbances, a higher rate of disturbance, and chronic, long-term problems affecting reefs, which make it more difficult for reefs to bounce back.

It's easier for corals to recover from an acute, short-term disturbance than from a chronic, long-term disturbance.[98] And chronic problems—like poor water quality and overfishing—combined with acute disturbances appear to make recovery particularly difficult, which is likely why Caribbean reefs were less able to bounce back from the coral and urchin diseases decades ago. The ability of coral reefs to recover from a disruption is being tested today, and many reefs are not able to bounce back. Sometimes recovery happens, but often it's impossible and the macroalgae phase persists.

Warming water caused by climate change adds another major wrinkle into reef resilience worldwide. When seawater gets too hot, coral colonies bleach, which leaves those that survive weakened and particularly vulnerable to disease and death. The global average temperature in 2023 was 2.1°F (1.2°C) warmer than the twentieth-century average, the warmest year on record as I write this in 2024, although there will no doubt be warmer years in the near future as the climate continues to warm.[99]

Increasing carbon dioxide, the most prevalent of the greenhouse gases warming the climate, is also causing another inescapable problem for corals, even those in remote places—it makes it difficult for corals and other marine life to build their limestone skeletons.

Some carbon dioxide in the air is natural: there were about 284 parts per million (ppm) of carbon dioxide in our atmosphere when the *Beagle* anchored in the Cocos Keeling Islands in 1836, which was about the same amount as had been in the air when George Gauld was making nautical charts of the waters off the Florida Keys. It had been about the same amount for centuries, give or take 10 ppm.[100] But the level has increased dramatically as we've burned fossil fuels: in 2023 there were 419.3 ppm of carbon dioxide in our atmosphere.[101] Some of that carbon dioxide dissolves into the ocean, causing seawater to become slightly acidic, which poses a threat to coral reefs worldwide, as well as to other marine life that builds skeletons and shells out of limestone. There's evidence that this ocean acidification makes coral reproduction more challenging too.[102] Early life stages of corals are particularly vulnerable. Acidification reduces the number of coral larvae that settle down into a reef over time,[103] which may slow recovery on reefs after disturbance.[104]

According to a recent report, if carbon dioxide and temperatures rise following the worst-case scenario modeled by the Intergovernmental Panel on Climate Change (IPCC), the group organized by the United Nations to assess the science of climate change, by 2034 all reefs worldwide will experience severe bleaching every year, and by 2050 nearly all reefs will be eroding rather than growing.[105] Climate change is also fueling stronger and wetter hurricanes, which can make a hurricane strike more devastating for reefs than in the past. However, there is hope: if we can limit emissions and keep climate warming below 2°C, nearly two-thirds of coral reefs may still be growing, building limestone, in the year 2100.[106]

We don't really know the relative importance of each acute disturbance and chronic problem that reefs face, although we do know that some compound the effects of others. But what's clear is that coral reefs are facing too many problems—local and global, acute and chronic. If we can tackle these problems, we will help increase reef resilience so that when reefs fall apart, it's only temporary.

Algae-covered and increasingly vulnerable reefs in the Caribbean today are not the key to the past, but their history can help us understand what happens when things fall apart. Their history also shows us what a healthy coral reef was like long ago, before we were watching them closely, which can provide a blueprint for how we can help reefs bounce back when they aren't able to do so on their own.

In order to help define goals for reef conservation and restoration, Katie Cramer and Loren McClenachan, both of whom we met earlier in the chapter, are now leading a group of scientists and conservationists to establish what a healthy reef should look like, using data from past reefs to inform reef management now and in the future.[107] If we can establish what reefs were like before humans started to change them, we can take steps to restore them to that baseline, with actions like protecting fisheries and adding corals to reefs that are grown in nurseries (which we'll visit in chapter 8).

While looking at evidence from decades to hundreds of years ago can help us see what went wrong, evidence from deeper in geologic time can give us perspective too. Looking further into the past can help us see how coral reefs have dealt with challenges at a larger scale than the disturbances we've seen here. Fossil evidence of how reefs have persisted over the long term through these challenges may help us better understand what it will take for reefs that have been decimated to someday return.

## CHAPTER 6

# WHEN REEFS PERSIST

It's fiercely sunny and hot on our first day at Devil's Point. Even the fossils are warm. Like many limestone-fringed coasts, this one on the Bahamian island of Great Inagua is an ancient reef. Some of the reef limestone is a creamy beige. Some is dark gray, its surface weathered by phytokarst.

Sally Walker darts from one spot on Devil's Point to another, joyfully exclaiming Latin species names when she spots fossils in the outcrop as if she's meeting up with old friends. Her husband, Fausto, follows her with an umbrella, attempting to keep Sally in its shade. Often Fausto is a few steps behind Sally, unable to predict her path through the outcrop. It can be hard to know what fossil she'll make a beeline for next.

Several steps behind Sally and Fausto, I'm walking the fossil reef with Al Curran and Mark Wilson, who are pointing out the evidence preserved in rocks and fossils that they and Brian White found at Devil's Point, which tell an unexpected story of what happened to Bahamian reefs during the last interglacial period: a brief drop in sea level killed shallow reef corals that were exposed above water. When sea level rose again, new corals grew in the same place.[1]

This was a place where a reef struggled and yet persisted. We'll get to know several such places in this chapter, including fossil reefs where corals died as sea level fell or rose too fast or when climate became too variable. Even though these prehistoric crises were very different from the struggles reefs face today, the fossil evidence can hint at whether modern reefs that are struggling could someday return.

In the world today, it's becoming increasingly common for reefs to not recover from disturbances—to not bounce back after a bleaching event or hurricane, for example. They have low resilience, likely because of a confluence of factors that we explored in the last chapter. But over longer timescales (although still short as compared with the vast expanse of geologic time), the fossil record contains numerous examples of how the ecosystem has been able to persist by rebuilding, usually with the same coral species and reef zones, after a time of struggle when the corals of a previous reef were killed. One example of these long-ago rebuilt reefs is preserved in the rocks at Devil's Point.

Devil's Point is a rocky headland on the west side of the large, remote island of Great Inagua, the southernmost Bahamian island, about 200 miles (320 km) south of San Salvador. Great Inagua is so far south that one can see Cuba from the top of the lighthouse on a clear day. Few tourists wind up on Great Inagua: of the more than six million people who visit the Bahamas in a typical year, only about a thousand visit Great Inagua. Our little crew of paleontologists may be the only visitors on the island this week.

Great Inagua is better known for flamingos and salt than for fossils. The epicenter of both flamingo and salt activity is Lake Rosa, a lagoon that fills much of the west half of the island. Morton Salt extracts about a million pounds of salt each year from Lake Rosa, harnessing the power of natural evaporation in a dry, hot climate to mine salt from seawater. We would see the results of this process, glimmering white mountains of salt, beside the lake.

Lake Rosa is also home to somewhere between 50,000 and 80,000 West Indian flamingos, the national bird of the Bahamas. They can be found wading through the extrasalty waters searching for brine shrimp, which give the leggy birds their pink color thanks to pigments in the algae they eat. Nearly 200 other bird species spend at least part of the year on Great Inagua, including colorful Inagua parrots that squawk from trees, burrowing owls that dart across the dusty ground, and hummingbirds that buzz flowers. Since 1965, more than half the island has been protected as Inagua National Park.[2] Henry Nixon, the park warden, is our main point of contact on the island. He meets the five of us at Great

Inagua's small airport. We hop in his pickup truck with our luggage and head to the guesthouse in Matthew Town, the island's only village.

It's January 1999 and this is my first time at the Devil's Point fossil reef on Great Inagua. It will become one of my research locations as I work on my dissertation. Eventually, I'll spend enough time on the island that locals will refer to me as "the Devil's Point lady," a moniker that makes me sound as if I'm the devil's apprentice, a go-to person for satanic activity, rather than someone who spends her days identifying fossil clams and snails preserved in an ancient reef.

As we walk Devil's Point, Al and Mark point out that what looks like one fossil reef is actually a stack of two reefs: a younger reef sits atop an older one. We stop every few steps, and they identify which lumpy fossil outcrops are from the older reef and which are from the younger reef. In many places, the top of the older reef forms a broad, flat area, which is the easiest place to walk through the outcrop, and underfoot it's clear how a brief wiggle of sea level eroded the top of the older reef. Looking down as I walk, I see fossil corals that were planed off as sea level fell and their tops eroded. Bulbous corals with their tops sliced off look like circles and ovals. A large brain coral nearby, its shape reduced to a plane, looks like one huge circle. This surface, a plane between the two layers of coral reef, marks a time of disruption. Al mentions that a similar erosion surface appears in the Cockburn Town fossil reef on San Salvador, which they found before they identified the surface we're walking on at Devil's Point.

Splotches of red-brown caliche, a component of ancient soil, are on the surface, although it is weathered away in some areas and blackened with phytokarst in others. Caliche forms on land when limestone dissolves in rainwater and then solidifies into a crust within or just below soil. Caliche on the planed-off corals implies that the fossil reef was exposed above sea level on land after its top eroded away. On top of the ancient soil and erosion surface sit hillocks of a younger Pleistocene fossil reef, evidence that after the place was exposed above water, it was eventually underwater again and a new reef formed.[3]

This eroded surface is an unconformity, much like the one that James Hutton peered into in Scotland at Siccar Point, which we visited in chapter 2. He saw a vast expanse of geologic time in that unconformity. The unconformity at Devil's Point doesn't hold such vastness. The thin line separating the older and younger reefs likely represents about a thousand years,[4] long enough for soils to form on the surface and plants to grow, but given that it took time for sea level to drop and then rise, it may have been completely above sea level for only a few hundred years. This unconformity at Devil's Point formed in the blink of an eye compared with the unconformity at Siccar Point, which formed over millions of years. It's more like a page missing from a book rather than an entire chapter. But like all stories with missing parts, it's a mystery that begs for explanation. What exactly happened?

While some evidence of what happened was eroded away at the unconformity, clues are preserved, including trace fossils of plant roots that skate across the surface. They must have grown down through soil, now long gone, before hitting the solid limestone of the older fossil reef and turning to grow laterally along it.

Based on the fossil clues, Al, Mark, and Brian determined that a brief blip in sea level caused the corals of the older reef to die, their tops to erode, and soil and caliche to form on top of the limestone skeletons, enough for plants to grow. When sea level swung back upward, the younger reef grew atop the older.[5] The ecosystem was able to persist in the long term.

As we walk across the unconformity, Mark stoops over and points out small borings, burrows, and fossils from life that encrusted on the surface. His paleontology research often focuses on animals and algae that lived on and within rocky marine environments called hardgrounds. Mark is prone to get excited about the holes drilled into rocks by invertebrates, and the traces of life that encrusted over rocks, which are like the boring and encrusting life I found on the coral death assemblage from Florida Keys reefs. These traces stretch across the unconformity.

According to Mark, marine hardgrounds present a basic problem for all organisms: they need to avoid tumbling away. If you're going to live on rock, you have to either stick to it or drill into it, he tells me when we catch up over Zoom and discuss his research. Unlike a typical jumble of fossils in a layer of rock, the fossils and trace fossils on a hardground preserve a specific timeline of who lived when. "When you look at a hardground, these things that are encrusting are in their original position relative to each other," says Mark. "You can see the chronology of encrusting and there is an order to it. There are certain organisms that are there first, and then second, and then third, and then fourth."

The fossils preserved on a hardground make it possible to see how life in a place changed over time. This is a level of detail usually not seen in fossil-filled layers of rock, where fossils of animals that lived at slightly different times are mixed together. Mixing of fossils is called *time averaging* because the resulting bed of fossils shows an average of what lived there over time.[6] The shells on a beach, for example, may have belonged to snails and clams that lived over many years and their remains were mixed together. Time averaging makes it hard to see details.

Because the fossils on a hardground preserve the order in which species lived in a place with their layering, they can be a window into ecological succession, which occurs when species replace each other over time until, theoretically, they reach some type of climax community that is unlikely to change unless there is a disturbance. Ecological succession happens in many different types of ecosystems. A classic example is a forest in which the types of trees and other plants change over time as the forest matures. Imagine an area with exposed bare soil after the vegetation was burned in a wildfire or cleared away by a landslide. The exposed land may gradually become home to grasses, and then bushes and trees that thrive on sunshine move in. The trees may grow to be so dense that their young seedlings can't get enough sunshine, which allows trees that thrive in shade to move in. The process can take decades to thousands of years, depending on the ecosystem and whether the process is starting from scratch.

In fossil hardgrounds in the Bahamas, boring sponges tend to be the first life to show up, says Mark. They eat away at the limestone, which has left the unconformity at Devil's Point covered with tiny craters. Then clams bore into the limestone hardground, and small corals encrust on top of it, flat like pancakes. At this point, worms sometimes live on the surface, secreting suits of limestone that attach them to the rock.

Eventually, larger corals colonize the surface, growing the reef's limestone structure on top of the worm tubes, small corals, and sponge and clam borings.

Here in the Bahamas, Mark found a particular sequence of understated fossils in the unconformity between the older and younger Pleistocene reefs that has been key evidence explaining what happened before, during, and after this unconformity formed at some point between 125,000 and 120,000 years ago.[7] At the center of the evidence are fossilized coralclams.

Coralclams (*Coralliophaga coralliophaga*) live nestled in holes drilled by other animals. They can't drill their own holes into limestone so they rely on hand-me-downs. Once ensconced, they are protected from predators, and the coralclams grow to the size of their holes.

In the fossil reefs on both San Salvador and Great Inagua, Mark found fossil coralclams nestled in holes that had been drilled into the limestone of the older fossil reef by mollusks called date mussels (species of *Lithophaga*), which are adapted to use their body like a drill to plow through rock.[8] Since date mussels live in the ocean, the holes must have been drilled underwater. The tops of these holes are planed off, evidence that the date mussels lived before the top of the older fossil reef was eroded as sea level fell. Ancient soil—caliche and soil sediment—lines the inside of the date mussel holes, evidence that soil was forming on top of the fossil reef when sea level was lower and the reef corals died. The coralclams then moved into the soil-lined holes. They, too, live in the ocean, so they could have taken up residence only once sea level rose again. These fossil "turduckens"—a clam inside soil inside a hole drilled by another clam inside a coral skeleton—preserve a huge story of regional or global sea level change.

It was at the Cockburn Town fossil reef on San Salvador that Mark first found the coralclams and evidence around them. This was in the 1990s, after the Bahamian government blew up part of the Cockburn Town fossil reef to create a small marina on the north end of town, which exposed fossils that had been trapped within rock for over 100,000 years, including evidence that two reefs formed—one after another—during the last interglacial. Before Mark explored the fossils exposed by the blasting, the Bahamas fossil reefs on both San Salvador and Great Inagua were assumed to have formed at one time.

"My thought was, well, since they blew it up then at least we could go look inside it so we could see what this reef looks like," Mark recalls. "I dropped down into this hole and just saw amazing things."

The dynamite had exposed the erosion surface. What looked like just a line in cross section was exposed as an eroded plane at the top of the older reef. There was a clear unconformity between two fossil reefs and there was a world of fossil life on the unconformity—tiny encrusting corals, lots of sponge borings, and best of all, coralclams within soil-lined date mussel borings.

"I did immediately know that this was a surface of erosion that had been occupied by soil—by dry land—for some period," says Mark. There was evidence that sea level dropped far enough for plants to grow on the eroded top of the older reef and evidence that sea level rose again. Eventually sea level rose enough that corals started growing on top of the bored and encrusted older reef.[9] Mark rushed back to the Gerace Research Centre to tell Al and Brian what he had found in the hole. "I came back and I was so excited. I said, 'Al, Brian, you're not gonna believe this!' And they did not believe it," Mark recounts. Al and Brian were skeptical. But that just energized Mark to search for more evidence in the fossils. Eventually all three collaborators would document the erosion surface within the fossil reefs of both San Salvador and Great Inagua.

At an even smaller scale than a coralclam, they found evidence of ancient sea level change that's visible only with a microscope, preserved in the spaces between sand grains and fossils within the reef limestone. Sedimentary rocks, including limestone, are cemented together with mineral crystals that grow between the sediment and fossils, binding them together. On limestone platforms, these mineral crystals are types of calcium carbonate that grow as water filters between the sediments. Different types of calcium carbonate minerals grow from fresh water and salt water, which preserves a record of whether the rocks have been below sea level, above it, or both. The mineral crystals that formed in the older reef at Devil's Point are found in three layers—the oldest layer of crystals indicates the reef was first cemented together on the seafloor;

**Two reefs built during the last interglacial**
Cockburn Town fossil reef, San Salvador Island, Bahamas

the second layer formed when the reef was exposed above sea level and affected by rainwater instead of seawater; and the third layer of crystals indicates that the reef limestone was eventually back underwater.[10]

If you were a coral polyp living in a colony attached to the seafloor and sea level fell enough to leave you above water, you would die. There is nothing particularly resilient about that. You would have no way to evacuate. But your species, and reefs as ecosystems, have ways of persisting.

The lumpy limestone deposits that sit atop the unconformity at Devil's Point are the remains of a younger reef that formed once sea level was high again. In some places only a jumble of coral rubble is preserved from this younger fossil reef, including branches of staghorn coral mixed with mollusk shells and seafloor sands. In other places, large heads of coral sit on the unconformity right where they grew. Spaces between the large coral heads are filled with branches of staghorn coral and sand.

After sea level rose again in the late Pleistocene, the unconformity was a new hardground in shallow tropical water, just the place for reef corals to grow. For this to happen, corals would need to get there. Corals are sessile (stuck in place), so they can't scamper to a new location. But they do have an intergenerational way of getting around. Most corals start their lives as larvae floating in seawater as part of the zooplankton. The longer they float in the water before settling down on a hard surface, the farther they might travel. Some coral larvae spend just two or three days floating in seawater before they settle down.[11] Others can float around for more than 10 weeks.[12] Once a coral larva settles down, it metamorphoses and starts to construct its limestone skeleton. At Devil's Point, the coral larvae settled down and started to build their skeletons on the sheared-off tops of corals that had died about a thousand years before.

The wiggle in sea level that decimated the reefs at Devil's Point and Cockburn Town and then allowed them to rebuild was minor compared to the wild swings of sea level that happened as glaciers and ice sheets formed and melted repeatedly through the past 2.6 million years of the ice age.

Before human-caused climate change, cycles of cooling and warming during this ice age were controlled entirely by changes to our planet's axis of rotation and the shape of its orbit around the Sun. For about the past 800,000 years, these waves of cool and warm have happened about every 100,000 years.[13] As a result of the cycles, ice sheets and glaciers, most near the poles, would grow, melt, and grow again, over and over.

Nearly all the ice was as geographically distant from coral reefs as it's possible to be, yet it had an enormous effect on coral reefs. When ice sheets and glaciers are smaller, during interglacial periods, more water is in the ocean and sea level is higher. When ice sheets and glaciers are large, during glacial periods, there's less water in the ocean and sea level is lower. Additionally, when water warms, it expands, which also raises sea level. When it cools, water contracts, which lowers sea level. About half of the recent sea level rise is from melting ice and half is from warming water.

Fossil coral reefs preserve a record of where sea level used to be because their shallowest parts were typically just below sea level. When we find fossil coral species that prefer the shallow reef crest, we can assume that the water was relatively shallow in those areas. During the last interglacial period, climate was warm and sea level was high, which is why these fossil reefs are now mostly on land (at least in places where land is not sinking). But, as the Bahamian fossil reefs and their embedded unconformity illustrate, sea level wasn't entirely steady during the last interglacial period.

Many fossil reefs around the world also preserve similar evidence of shifts in sea level during the last interglacial period, but curiously, reefs in different locations seem to have been affected differently by sea level changes.[14] In the reefs at Devil's Point and Cockburn Town in the Bahamas, there's evidence the older reef formed when sea level was about 13 feet (4 m) above the present level, dropped about 13 feet, just barely exposing the older reef, and then rose to 20 feet (6 m) above the present level.[15] Fossils from Treasure Beach, Jamaica, show two different periods of reef building during the last interglacial, with a calm-water patch reef atop an older reef filled with elkhorn coral.[16] In Yucatán fossil reefs, there's evidence of two peaks in sea level during the last interglacial of a similar magnitude to that in the Bahamas, but no evidence of a sea level fall between the peaks.[17] In Australia, there is also evidence of two peaks in sea level, with the second peak 31 feet (9.5 m) higher than sea level today, far more sea level rise than at other locations.[18] Fossil reefs near the Red Sea record at least four wiggles of sea level during the last interglacial period.[19]

Because different reefs tell different stories about sea level during the last interglacial, it's challenging to know what was happening globally. However, what is clear is that reefs rebuilt each time sea level changed.

Changes in sea level of a magnitude that affected reefs imply that ice sheets were changing at the time. But changes in the ice can create another effect: they can change the level of the Earth's crust. When glaciers and ice sheets melt, the Earth's crust underneath, suddenly free of the ice that weighed it down, rebounds, slowly regaining its shape like a cushion on a couch that's recently been vacated. And when the crust rises, sea level shallows in that particular location, which could explain regional differences in fossil reefs from the last interglacial.[20]

We're in an interglacial period now, which makes the question of how sea level jumped around during the previous interglacial period particularly timely. Coastal cities made by humans and shallow reefs made by corals are both vulnerable to even small changes in sea level. Several feet of sea level rise can flood a coastal city. Several feet of sea level fall can leave a reef exposed on land, which is what happened at Devil's Point. Understanding the sea level changes during the last interglacial could help both humans and reefs today.

If sea level keeps rising at the rate it's rising now (0.14 inch, or 3.6 mm per year),[21] the fossil reef at Devil's Point and many others will be underwater again in a thousand years, and so will many of our coastal communities. But there's evidence the rate will increase as climate warms, which would inundate coasts sooner. If the Greenland ice sheet melted or slid into the ocean entirely, which is possible but very unlikely, coastal communities would flood with a sea level rise of nearly 24 feet (7.2 m).[22] Living reefs on the seafloor would find themselves in much deeper water, and new reefs might form on top of the fossil reefs from the last interglacial period, preserving another unconformity between past and future reefs.

∼

As we've seen, falling sea levels can mean death for corals. When sea level fell a small amount in the Bahamas during the last interglacial, corals died, but the edges of the limestone platforms were still underwater at that time, so corals could live there. However, when sea level later dropped hundreds of feet as ice sheets grew and the glacial period set in, life in reefs could not persist anywhere on top of shallow limestone platforms or even the upper part of the platform sides. These parts of limestone platforms had become dry land, and all coral and other marine life on them died.

Perhaps counterintuitively, rising sea levels can also spell death for corals when the rise is particularly fast. As the last glacial period ended, ice melt wasn't always a steady, gradual process. Sea level jumped rapidly with pulses of meltwater, at times rising as much as 100 feet (30 m) in a thousand years.[23] With sea level low, reefs were on the sides of platforms. Many had trouble growing upward fast enough during rapid sea level rise. No matter how much limestone they were able to produce to build upward, the coral couldn't stay in shallow water.

If sea level rises faster than corals can build upward, and the reef winds up in water too deep for light to filter through, corals will die and the reef will perish. This is known as a *drowned reef*. Drowning is a somewhat peculiar term to apply to life on the seafloor. In this case it's a lack of sunlight that causes the drowning—without sunlight, the zooxanthellae in corals can't photosynthesize and corals can't survive.

About 11,300 years ago, a pulse of meltwater from shrinking ice sheets drowned reefs off the south coast of Barbados as sea level rose between 26 and 36 feet (8 and 11 m) in about 250 years.[24] Drowned reefs have also been found off the islands of Hawaii[25] and Tahiti.[26] Not all reefs drowned with pulses of meltwater; some with fast-growing coral species appear to have kept pace with sea level rise.[27]

Offshore from where Australia's Great Barrier Reef is located today, five fossil reefs are currently in water that's about 140 to 550 feet (42 to 167 m) deep. They lived from over 30,000 years ago to 10,000 years ago—the time leading up to, during, and just after the height of the last glacial period when glaciers and ice sheets were at their largest and sea level was at its lowest. The growth of all these reefs was eventually halted because of either rising or falling sea level.[28]

The International Ocean Discovery Program (IODP) extracted cores from these fossil reefs in 2010. The IODP, which was called the Deep Sea Drilling Project in its early years, has been drilling cores of seafloor rocks and sediments around the world since 1966 with a drill rig mounted on a research vessel.[29] They send the drilling bit below the ship, through the water, to extract cores from the seafloor.

When the IODP science team analyzed 20 cores from these five fossil reefs near Australia, they found that two of the reefs lived while glaciers and ice sheets were growing in the time leading up to the last glacial period. Both of these reefs eventually died as sea level fell. There was evidence in the cores that the reef's limestone was exposed above sea level and affected by rainfall. Three other reefs grew at times when sea level was rising, sometimes quite quickly, as glaciers melted and the glacial period ended. Each of these reefs drowned as sea level rose, but the team found that it wasn't sea level rise alone that caused the drowning. In part, the reefs drowned because coral growth slowed down as runoff from land reduced water quality. With sea level rising, slower-growing reefs couldn't keep up.[30] This same vulnerability of corals to sediment and lower water quality that we see in reefs today also affected these ancient reefs. Back then, the water quality declined because of erosion as the continent was flooded by rising sea level; today, water quality is most dramatically affected by Anthropocene activities such as coastal development and the use of chemical fertilizers, as we saw in the previous chapter.

These five reefs died, and yet the ecosystem as a whole persisted. There weren't large gaps in time when reefs weren't growing in the region. While one reef was dying, or soon after (often within a thousand years, which is a long time for us but short geologically speaking), a new reef would start growing where water depth was optimal. Sometimes the next reef started growing at about the same time that the last reef was dying. The IODP science team also found that the corals persisted despite changes in water temperature. The chemistry of coral skeletons records a change in temperature of about 7°F–9°F (4°C–5°C) over 10,000 years, which implies that these corals had some resilience to changes in temperature.[31] However, water was warming at a much slower rate than it is today, so corals' past ability to cope might not indicate they will be able to survive Anthropocene climate change, which we'll explore in chapter 9.

While reefs worldwide bounced back through ice age swings in sea level, some vulnerable reef inhabitants didn't make it. Two coral species found in Caribbean Pleistocene fossil reefs became extinct at some point in the past 80,000 years, long before humans had much influence on reefs. Both may have gone extinct during the last glacial period, when low sea level meant that there were far fewer areas for Caribbean reefs to form.[32]

The organ pipe orbicella (*Orbicella nancyi*) had been abundant on Barbados reefs for at least half a million years before it disappeared, according to John Pandolfi.³³ *Pocillopora palmata*, the other now-extinct coral, formed upright curved blades of limestone that, in cross section in fossil reefs, look like collections of apostrophes, parentheses, and other curvy punctuation marks. It's found in fossils as young as 82,000 years old in Florida and Barbados and hasn't been seen since.³⁴ Given how much less area there was for reef building when sea level was low, it's amazing that all corals except for these two survived the ice age.

---

Reefs have also been affected by environmental changes in the past that have nothing to do with the ups and downs of sea level, and these changes can also create unconformities in reef limestone. Sea level was stable or rising gradually when the reefs off Panama's Pacific coast stopped growing about 4,000 years ago. After a 2,500-year hiatus, they started growing again.³⁵ This wasn't a global event. As far as we know, only reefs in the Eastern Tropical Pacific were affected, a region that stretches from the west coast of Mexico south to Colombia and the Galápagos.

Coral reef ecologist and geologist Lauren Toth and her collaborators studied the hiatus in reef building to figure out why it happened. Underwater, they extracted cores from below the seafloor by banging a metal tube into the reef rubble, a method similar to how Katie Cramer's group extracted cores from below reefs in Panama and Belize in search of parrotfish teeth. Within those cores, they found reef corals that grew until a little over 4,000 years ago. Above those corals was an unconformity in the cores, a gap in reef growth (although no actual gap in the cores). Above the unconformity were corals from a younger reef that started building limestone about 1,600 years ago and still persists today.³⁶

Unlike the unconformity between the older and younger reefs that Mark and Al showed me on Great Inagua, this much younger unconformity off the coast of Panama did not show signs that fossils and rocks had eroded away. An unconformity does not have to involve extensive erosion. It can instead form during a time when no sediments or fossils are deposited. In this case, nearly nothing was preserved for two and a half millennia. Lauren and her team set out to investigate what types of environmental change could have shut down the reef. They would come to find that because this area has challenging conditions for corals, when the environment became a bit more inhospitable, reef building stopped.

While corals have very specific environmental needs—such as clear water low in nutrients and a limited temperature range—some reefs make do with less-than-ideal conditions, including those of Panama's Pacific coast. Reefs living on the edge of what corals can tolerate are known as *marginal reefs*. Most coral reef research has focused on reefs in ideal conditions with clear water and consistent temperatures, so less is known about marginal reefs. Yet they may hold valuable clues to what it takes to survive.

These marginal reefs in the Eastern Tropical Pacific are generally small and have lower coral diversity than other locations.[37] There are fewer coral species at least in part because the reefs are separated from the high-diversity Pacific reefs to the west by a wide stretch of deep ocean, and currents don't bring coral larvae into the area. But the reefs are also small because environmental conditions are not ideal. Water temperature varies and the water is more acidic. In some areas, the water gets clouded with sediment.[38] It wouldn't take a huge shift in the environment for reef building to halt.

Lauren's research has a particular focus on marginal reefs. She investigates what they were able to tolerate in the past and what conditions have made them unable to cope. By looking at marginal reef fossils and subfossils, she can see how these reefs changed over thousands of years, longer timescales than she could possibly explore by monitoring living reefs. She and her team postulate that reefs stopped growing in the Eastern Tropical Pacific for 2,500 years because of swings in sea surface temperature associated with the El Niño–Southern Oscillation (ENSO), a cycle of climate that includes two extremes: El Niño and La Niña.[39] While ENSO has worldwide effects, the main changes are in the Pacific. During El Niño, air pressure drops in the eastern Pacific and rises in the western Pacific, which causes the trade winds to weaken. With weaker trade winds, less cool water upwells from the deep near the coast of South America, and a tongue of warm water spreads across the tropical Pacific. This El Niño phase can put corals off the coast of Panama into hot water. During La Niña, the opposite occurs—trade winds become stronger across the Pacific, which increases upwelling, putting the corals off the coast of Panama into colder water brought up from the deep.

ENSO is an ongoing process, but it hasn't always had the same intensity. There's paleoclimate evidence that the El Niño and La Niña phases may have been stronger and more variable during the time when Panama reefs shut down. If so, corals may have died during wild swings in water temperature. It's also possible that, weakened by changes in temperature, corals were too sensitive to cope with other types of environmental change.[40]

Knowing that limestone wasn't produced at this location for more than two millennia makes me curious about what was happening on the seafloor at that time. Did everything in the reef die? Was there nothing living on the seafloor? Was there life there that wasn't preserved? If so, that's another page missing from the story.

I ask Lauren whether there is a way to know what was going on at that time, and she tells me that there are a few clues as to what was living on the seafloor. Although not much of anything was preserved, there are some small corals from that time within the cores, evidence that there was still reef life in the area even if reefs weren't actively building. "That gives us some hints that these local populations of corals were surviving," she explains. "In the geologic past, a lot of times when these reefs weren't building new structure, growing, there still probably was a reasonable amount of living coral." It's likely that there was still a reef ecosystem even if only tiny traces of it are left.

If a reef shuts down geologically, it's not producing much limestone over time, as described in chapter 1. Reefs that are growing in a geologic sense create a certain amount of limestone each year. Some of that limestone is destroyed by waves and chewed apart by bioeroders. It's like limestone accounting. Growing reefs are in the black, producing more limestone than is eroded away. When reefs shut down geologically, their limestone production is in the red. If a reef doesn't produce at least as much limestone as is eroded away, it will shrink over time, be at risk of drowning, and won't leave fossils behind. That's what many reefs are facing in the Anthropocene.[41]

The reefs off the west coast of Panama were geologically shut down, but the seafloor may have still looked reef-like in some places because a reef that has geologically shut down can still function ecologically. The ecological community drives limestone production, but it can go on even without limestone building. In some cases, an environmental change causes a reef to shut down both geologically and ecologically, as when sea level fall leaves a reef above the water, as it did on Great Inagua, or when sea level rise leaves a reef too deep for corals to survive, as happened to the ancient Great Barrier Reef when glaciers melted. Panama's Pacific reefs may have shut down only geologically, but there may have been living corals on the seafloor, at least in the more sheltered areas. There may have been reef fish and invertebrates. It could even have looked like a coral reef, Lauren tells me, but if a reef isn't building limestone over decades to millennia, then geologically it's not functioning.

What was happening on Eastern Tropical Pacific reefs during the 2,500-year gap may have been somewhat like what is happening in Florida today—an example of how uniformitarianism can, in some cases, still apply in the Anthropocene. In Florida, reefs are no longer producing much if any limestone, yet there is still reef life on the seafloor. According to Lauren, many parts of the shallow seafloor off the Florida Keys are made of limestone that formed 3,000 years ago. Newer limestone has not been deposited above it. That shallow seafloor is an unconformity in the making.

With an interdisciplinary team of geologists, oceanographers, and ecologists, Lauren tracked the expansion and contraction of Florida's reefs through time using cores from below the seafloor unconformity. The team found that reefs grew fairly rapidly around 7,000 years ago, producing limestone about a foot (30 cm) thick every 100 years.[42] Then the rate of growth declined dramatically, and by about 3,000 years ago most reefs in Florida had shut down geologically.[43] This means that the reefs mentioned in chapter 4 where I collected the coral death assemblage in Florida had been shut down geologically for thousands of years. I had no idea at the time. They still had about 30 percent living coral and were functioning ecologically, although that wouldn't last. Today, most Florida reefs, having lost living coral, are now losing limestone.[44]

Today, reefs off the Florida Keys face multiple Anthropocene stressors that are causing corals to decline, including our warming climate. But Lauren thinks the reef limestone production slowed several thousand years ago because of a type of climate change we don't often think about today: cooling.

Florida's reefs are in the subtropics, at the fringe of water warm enough for reefs to survive. This is another marginal environment for reefs. The reefs are so far north of the equator that the corals are sometimes subjected to cold in winter. But they are also bombarded by hot water in the summer and fall, which means they are at risk of both heat stress and cold stress at different times of the year.

Around 7,000 years ago, during a warm period known as the Holocene thermal maximum, sea surface temperatures in Florida were probably slightly warmer than the twentieth-century average (yet not as warm as they are becoming now).[45] The Holocene thermal maximum occurred as glaciers melted, which caused more sunshine to be absorbed by the Earth instead of reflected out into space by the ice. Most of the change in temperature was near the poles, but there's evidence that the tropics did warm slightly.[46] Florida reefs thrived in the warmer conditions. During the last interglacial, Florida reefs had been limited to the Keys, the lowest and warmest latitudes in the state, but during the Holocene thermal maximum, warmth allowed reefs to extend farther north along Florida's east coast.[47]

When climate turned slightly cooler and more variable, 5,000 or 6,000 years ago, the rate of reef limestone production in Florida started to slow.[48] Other reefs throughout the Caribbean region were growing at this time, but Florida's reefs were struggling. "What's probably more important than cooling overall is these cold fronts that we periodically experience that aren't experienced in more tropical regions," notes Lauren. Cold stress events are rare in Florida today, but they do happen and they can kill corals faster than heat stress. In January 2010, cold weather caused shallow water in the Florida Keys to drop below 61°F (16°C), the lower limit of what corals can tolerate. In the weeks after the cold snap, scientists surveyed Florida reefs and found catastrophic impacts: on average, 11.5 percent of corals in Lower Keys reefs died in the cold.[49] Coral colonies that had lived for centuries died within days.[50]

If we had taken action decades ago and kept the amount of anthropogenic warming small, that might have brought Florida's reefs back into ideal conditions, but it's too late for that, says Lauren. Today there is too much hot water in the summer because of rapid human-caused climate change, and, because seasons happen each year, there are still winter storms.

Without corals actively producing limestone, Florida reefs are in the process of drowning. Many aren't keeping up with sea level rise, and with erosion from storms and bioeroders outpacing limestone production, many reefs are eroding away.[51] "Erosion is a slow process. It's not like you're going to be able to watch an entire reef crumble in a decade," Lauren tells me. Some of these reefs were built over thousands of years but could erode much more rapidly. "If the rates of erosion that we're seeing keep up or accelerate, it could just be a couple of centuries before all of that structure is lost," she adds.

However, given thousands of years and reef-friendly environmental conditions, it is entirely possible for reefs to return, growing on top of the unconformity, as long as the species don't become extinct. Reefs returned to the Eastern Tropical Pacific near Panama after a gap of only a couple of thousand years or so, which is brief compared to the entirety of geologic time, but it would feel like forever if you were there, watching and waiting.

Where did corals go during these breaks in reef growth? In the Panamanian Pacific, Lauren suspects that there may have been pockets where corals could have survived the difficult times before they repopulated the area. Off the coast of Australia, the shape of the seafloor, with its variety of shallow and deep habitats, may have provided enough safe spaces for corals to survive between periods of reef building around the last glacial maximum.[52] During times of environmental change when reef growth isn't possible, nearly all coral species and other reef life avoid extinction by retreating to safe places where they can persist. But where are these places? And are there enough safe places as we head into a warmer future?

## CHAPTER 7

# LYING LOW TO AVOID EXTINCTION

Corals are stuck in place, but they can relocate through their descendants—most have larvae that float in the ocean and settle down, growing colonies in new areas—which has made the history of reefs over glacial-interglacial cycles of the ice age like a game of Whac-A-Mole. When sea level changed, often by hundreds of feet, reefs would die and then a new reef would pop up elsewhere as coral larvae settled down at a better water depth. Tropical oceans stayed warm enough for reefs to form during recent ice age cycles, but changes in sea level meant that reefs kept finding themselves too deep or too shallow. Coral reefs grew, were decimated, and regrew over and over as sea level changed over time.[1]

Unlike in Whac-A-Mole, a new reef wouldn't necessarily pop up immediately. If you linked up all the fossil reefs we know about, they wouldn't form a continuous timeline documenting reefs forming at all times. Instead, there were times when reefs weren't actively growing. Despite the dearth of reefs at times, corals and other reef species didn't go extinct for the most part. Between the times when reefs were growing, corals and other reef life needed to find safe places to survive.[2] These places might not have been ideal, just tolerable, like Covid lockdown in a tiny apartment. Reefs might not have been able to actively grow in such places, but the species could lie low and wait out tough conditions.

Places where species can persist during a crisis are known as *refugia*. Refugia may be small and isolated areas compared with the range of a species in better times. They can be particular latitudes or habitats where species can find livable conditions.[3] Once the crisis ends, species can emerge from refugia and expand into more areas,[4] just as we eventually emerged from Covid lockdown and expanded into places we had abandoned like indoor restaurants and movie theaters. Unlike with the Covid lockdown, species typically shelter in refugia for generations until environmental conditions improve.[5]

It's thanks to refugia that corals had safe places to survive in the past. And refugia will hopefully be able to offer the same protections as we head into a warmer future, although it's uncertain whether there will be enough safe spots for all coral species to survive.

Refugia can be defined at different scales. Some consider refugia to be at a large scale: one large location that supplies a whole region with life when conditions improve. Others think of refugia at a smaller scale: many little pockets where species can stay safe. Over time, refugia have helped species from many different ecosystems survive. On land, surviving during the ice age often involved changes in latitude, especially in northern regions. During glacial times, animals and plants (via the spread of seeds) would move south to lower-latitude refugia to avoid ice sheets and extreme cold. During interglacial times, refugia at high latitudes became home to species adapted to cold when lower latitudes were too warm. There were also small refugia, such as microclimates in landscapes, where species could find tolerable conditions.[6]

Today, scientists are on the lookout for refugia that could shelter wildlife from anthropogenic climate change. If we know where these locations are, we can help protect them from other types of human-caused destruction. This is not a long-term solution—we need to address climate change—but refugia could help species make it through the decades to come as temperatures climb.

Refugia are often hard to find in the fossil record because they are typically small. One clue that there must have been refugia is a species that is fossilized in rock that formed at a particular time, then disappears from the fossil record and reappears at some later time. Species that follow this pattern are called *Lazarus taxa*, referring to the biblical story of Lazarus being temporarily dead, because they seem to be temporarily extinct.[7] But they were not really temporarily extinct. We just can't find their fossils. They must have been somewhere. It's possible that Lazarus taxa were lying low in refugia during the time when they can't be found in the fossil record, and those refugia have not been preserved. Or maybe they just haven't been found yet.

Glacial periods—when glaciers were large and sea level was low—posed the greatest challenge for coral reefs during the ice age because there were far fewer spaces where reefs could grow.[8] About 20,000 years ago, at the height of the last glacial period, sea level was about 420 feet (130 m) lower than it is today, and the tops of limestone platforms were high and dry.[9] Spaces for reef building that had been available during the previous interglacial were no longer an option. Had the technology existed back then, satellite photos of tropical oceans would have had a lot less light blue color. The amount of shallow ocean shrank because, in many areas, sea level fell below the nearly flat limestone shelves that are flooded with shallow water during interglacial times. There was about 90 percent less shallow ocean during the last glacial period in most areas of the tropics, except for Hawaii, where there was 75 percent less.[10] The Pacific and Indian Oceans became somewhat separated because, with sea level low, there was far more dry land between Papua New Guinea, Indonesia, the Philippines, and the Asian continent.[11] The Red Sea was cut off from the rest of the ocean as sea level fell. It would have become extrasalty because of evaporation, so no coral could survive. (This makes it particularly impressive that the same coral species we see in fossil reefs are also living in the Red Sea today, which we'll explore in chapter 9).[12]

If you were in a boat sailing around limestone platforms 20,000 years ago, you would pass high islands with tall, steep sides and flat tops, like mesas of limestone.[13] You wouldn't find many harbors, just rocky cliffs and landslide debris where the land met the sea at the platform sides.

Waves would be crashing against the coast. Those waves could have been especially powerful since there wasn't much shallow water to slow them down. With sea level low, reef species would have been able to survive only on the steep platform sides.

Limestone platforms, as you may recall from chapter 1, typically build over time like a layer cake with new layers of limestone added to the top. On atolls, the limestone platform is atop a sinking volcano, as Charles Darwin discovered. During interglacial periods, most of the top of the platform is in shallow water where corals, algae, and other organisms create limestone. But during glacial times when sea level falls, the top of the platform is above sea level. There's no more limestone production up there. The ocean surface meets the side of the platform at cliffs, piles of rocks that have cleaved off, and aprons of sediment.[14] The sides of a limestone platform are not usually smooth and uniform but instead can have stair-step terraces, cracks, overhangs, and scars of erosion.[15]

The state of a platform's sides depends in part on whether it is made of limestone mud, sand, or corals.[16] Parts made of corals often form steep or even vertical walls of limestone because corals stick together like Lego® bricks, forming a solid blocky object that can have nearly vertical sides. But loose sediment is like a pile of tiny detached Lego bricks—prone to scatter all over—so it is often less steep.[17] Some platform or atoll sides are vertical and others are more like the slope of a ski run. Either way, eroding sediment and coastal cliffs are not a great place for reefs to form.

If you stood on the coast of San Salvador, Bahamas, during the last glacial maximum and looked out to sea, the tropical Rothko painting would likely have two stripes of blue instead of three: just the deep water and sky. If there was a shallow turquoise stripe, it would be very narrow, located at the base of the cliff that you were standing on at the edge of the mesa-shaped island. Less shallow turquoise water means far fewer shallow sunlit areas where corals could grow.[18] The flat top, a shallow playground for reef life during interglacials, would be about 30 stories above the ocean surface. A coral larva floating around would find fewer places to settle down and grow and would probably wind up at the side of the platform.

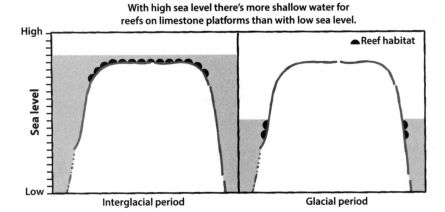

**Reef habitat depends on sea level**

With high sea level there's more shallow water for reefs on limestone platforms than with low sea level.

"They obviously persisted because the same species came back with the rising sea level," says Al Curran, who introduced me to the fossil reefs on San Salvador and Great Inagua. I'm envisioning the struggles that shallow corals must have faced on the side of a platform during the last glacial maximum, but Al disagrees. "They went through that huge physical change with no problem," he tells me.

"Don't think of corals as being fragile out there," says Al, aware that I'm feeling sorry for those poor corals that lived 20,000 years ago. "Actually, they're extremely resilient."

Given that all but two coral species made it through the last glacial period without going extinct,[19] there must have been parts of the ocean where reef life could survive tough times. There were some areas where large reefs could keep building, like the ancient reefs off the coast of Australia described in chapter 6, which formed and re-formed as sea level changed,[20] but in many areas, conditions weren't ideal. Corals needed to find places where they could survive and wait out the crisis. They needed to shelter in refugia.

Areas that would have been too steep for reefs to form could still have been refugia for corals, according to Paul Blanchon, a geoscientist who researches how reefs formed and died during ice age sea level changes. "You'll probably find, in supposedly inhospitable areas where reefs didn't develop, coral still grew there," he says.

As with the scattered corals that Lauren Toth found in her research at the unconformity in Eastern Tropical Pacific reefs,[21] there could have been ecological activity, but if the corals in refugia were not building much limestone, they might not have left many fossils behind. "Really, corals will grow on anything that's firm and won't roll around, so the fact that you don't have reef structures doesn't necessarily mean you don't have corals," Paul adds.

There's still a lot we don't know about coral reef refugia during the last glacial period because the fossil evidence is now in deep water.[22] Cores have been taken of the seafloor that capture evidence of fossil corals, but we don't have the full story of where corals took refuge. I love a good mystery, and the mystery of missing refugia ignites my curiosity just as much as the missing stories at unconformities, so I can't help but wonder what the life of a coral would have been like in a refugium when sea level was low.

Today we can dive about 100 feet (30 m) down to see what life is like on the upper part of a limestone platform's sides, which can give us a clue to what it might have been like for reef life surviving on a steep slope during the last glacial period. A dive to the side of a limestone platform is called a wall dive, and it will typically take you to the forereef zone—the zone seaward of the shallow reef crest, often on a steep slope, as we learned in chapter 3. During the last glacial period, the wall would likely have been home to not only forereef species but also species that, during interglacials, typically thrive in the quiet backreef zone and the shallow reef crest on top of a limestone platform. With sea level low and the platform top high and dry, those environments weren't available during glacial times. The reef life would have been in areas of the wall that today are at depths well beyond the reach of scuba. But we can get a sense of what wall life would have been like by exploring the modern forereef zone at a wall.

To set out for a wall dive, you'll often need a boat to get to the dive site at the platform edge unless that edge is quite close to shore. The boat will take you across the turquoise platform top to where it meets the dark blue. You'll be diving where the two colors meet at the edge of the abyss.

You'll splash into the water and signal to the people still aboard that you are okay, release the air in your BC (buoyancy compensator) vest, and drop down slowly to get to the wall, the regulator in your mouth bubbling with each exhale. As you go deeper, colors deplete. First red will disappear and then orange and yellow because at depth, sunlight strains to make it through the water. You'll still see some colors, but they will all have a blue cast. At some point, you'll no longer see the glimmers of the sea surface clearly, although it will still be lighter above so you'll know which way is up.

Wall dives are unique. Look down and you'll see the darkness of the abyss. Look to one side and there is a city of life clinging to steep or even vertical surfaces, cracks, and caves, perhaps with a swarm of fish nearby. In the deep parts of a reef, corals grow in flat plate shapes, their broad surfaces collecting the faint sunlight like solar panels for their symbiotic algae.

During the glacial period, all the round and branching corals that live in shallow reef zones today would have been on the steep slope. In this environment, it's helpful to be a species that cements itself in place, like corals and sponges. Soft corals like sea fans and sea whips, which don't create hard calcium carbonate skeletons, are also cemented in place on the wall. Their flexible branches bend in the current, but they are well anchored in place. A wall dive is a good reminder that lots of reef species are adapted for life on a vertical surface.

One reason I love wall dives is that I can be upright as if standing instead of horizontal as if swimming. I hover like a ghost in the water, and because there is nothing but water below, my fins will not hit corals. Dive with a group and you may find nearly everyone in a vertical hovering position looking intently at what's on the wall as if they're visiting an art museum. Sometimes there will be a current along a wall, so all the divers in your group will drift laterally as if you're all on a people mover, watching the wall of reef as you pass by.

While exploring life on a wall today can give us a sense of what it takes to survive on a vertical surface, it can't show us what life was like for corals in shallow water during the last glacial period. Unlike with a shallow reef, a wall dive today takes you to a habitat typically below the reach of waves. It's calm down there. But when sea level was low, corals living in shallow water must have dealt with waves smashing into them at the platform sides. If those waves broke coral branches, the fragments might have tumbled too deep to survive and start new colonies. There wouldn't have been the same protected areas behind the reef crest for corals that preferred calm, shallow water. Shrimp and other invertebrates that don't swim or cement themselves in place would have had to find nooks where they could shelter. Perhaps they retreated to deeper water to avoid the waves.

Erosion of the cliff would also have posed a hazard, as would sediment eroded by rivers emptying into the ocean, clouding the waters of nearshore reefs.[23] Murky water may not be ideal conditions for coral, but refugia often don't have ideal environmental conditions, just survivable ones.

While the locations of ancient reef refugia are hard to find, some animals alive today hold evidence of refugia in their genes. For example, researchers studying the genes of the coral *Acropora millepora* at 20 sites along the Great Barrier Reef found evidence of two refugia during the last glacial period, each with its own isolated population of the coral that had evolved differently to suit its refugium environment.[24] When populations are isolated long enough, they can evolve to become different because genetic traits in small, isolated populations can become magnified over time and because populations adapt to their environment. Given enough time, this produces different species. For example, finches, the birds that Darwin studied in the Galápagos, were isolated from each other for two to three million years and evolved different beaks. The *Acropora millepora* corals were isolated in refugia for far less time than the finches—long enough to accumulate different genetic signatures but not long enough to evolve into different species.[25]

Genetic analyses can also tell us when a species hasn't been able to get out of a refugium. This happened to blacknose sharks in the Bahamas after the last glacial period. The sharks took refuge in two locations when sea level was low: the southern Gulf of Mexico and the Bahamas.[26] The population of sharks in the Gulf of Mexico was able to emerge and expand throughout the region once sea level was higher, but the population in the Bahamas stayed in its safe place and is now genetically distinct from other blacknose sharks.[27]

Glacial periods weren't the only times when reef life needed refugia. There is also evidence that corals retreated into refugia during particularly warm interglacial times. About 125,000 years ago, during the last interglacial, fossil evidence shows that many reef coral species retreated away from the equator, keeping to somewhat higher, although still tropical, latitudes to avoid the heat. This left reefs at and near the equator with less diversity, which may happen again as human-caused climate change heats water and stresses corals.[28]

---

*Acropora*, a fast-growing genus of branching corals,[29] includes a multitude of species in the Pacific and Indian Oceans and two species in the Caribbean: elkhorn coral (*A. palmata*) and staghorn coral (*A. cervicornis*). There is also a hybrid between the two Caribbean species, *A. prolifera*, which is often called prolifera for short. *Acropora* corals flourished during ice age sea level changes, possibly because they grow quickly compared to other corals and so could often keep up with rapid sea level rise as glaciers melted.[30] They also have a history of being able to recover from disturbances like hurricanes because their broken fragments can start new colonies.

But worldwide, *Acropora* species have not shown the ability to bounce back from recent Anthropocene problems.[31] Both Caribbean *Acropora* species are now critically endangered.[32] In the Pacific, *Acropora* corals are a favorite food of the crown-of-thorns starfish. They are also vulnerable to warming ocean water and reduced water quality. If these stressors intensify over the next few decades as projected, *Acropora* species could eventually face extinction.[33]

However, there are refugia where Caribbean *Acropora* corals have survived the ravages of the past several decades. One of these, Coral Gardens Reef, in sun-drenched water off San Pedro, Belize,[34] has been home to both staghorn and elkhorn coral and also to *Acropora prolifera*, the hybrid between the two, as well as other coral species like boulder star coral. The large, dense thickets of staghorn coral stand out at Coral Gardens, making the reef look as if it's from the mid-twentieth century or even earlier, a time before white band disease and the decline in water quality.

Al Curran first learned about Coral Gardens Reef from locals in San Pedro in the late 1990s when he and marine ecologist Paulette Peckol were surveying Belize reefs as part of an effort to document reef health throughout the Caribbean and western Atlantic. Coral Gardens wasn't on their list to survey, but the large healthy stands of staghorn coral, a sight that had become uncommon, left an impression. Al introduced geoscientist and paleoecologist Lisa Greer to the reef, and the two have collaborated on research there ever since.

Lisa started a research program in 2011 focused on understanding why staghorn coral has thrived at Coral Gardens and what allowed it to be resilient, collecting data about the reef each year. "We've spent far too much time documenting things dying," says Lisa. The reef's staghorn coral has given her hope over the years that this species might be able to avoid extinction. That hope would eventually be dashed, but we'll get to that later.

Before she was focused on Coral Gardens, Lisa studied a fossil reef in the Dominican Republic where staghorn coral thrived for 4,000 years (which we visited in chapter 4). The first time she dropped into the shallow water and saw Coral Gardens, she realized she was looking at a type of reef that she had only heard stories about and seen fossilized—one with healthy thickets of staghorn coral. "It just took my breath away," she recalls. "This is what they used to look like. I thought, oh my gosh, this is what they're talking about."

By studying the subfossils below the living reef, Lisa, Al, and the rest of their team discovered that Coral Gardens had been a refugium for staghorn coral throughout the twentieth century, even as other reefs in the Caribbean floundered.[35]

Staghorn coral forms a thicket of limestone branches that fit together like chaotic scaffolding, with few right angles. Taking cores of the rickety limestone framework below the living reef was too risky. The structure could collapse like a soufflé. So the team, scuba diving with hammers and chisels, meticulously dug three pits, each a meter wide, to excavate the subfossil coral skeletons below, selecting locations that would not disturb living coral. To create a detailed chronology, they kept each five-centimeter layer of the staghorn scaffolding as a separate sample.[36]

They also collected additional samples of the staghorn coral death assemblage using a rubbish-grabber tool, a claw on a stick that can probe open areas within a staghorn canopy without touching live coral. I wish I had had a litter picker when I was collecting the Florida coral death assemblage. That would have helped me win the underwater game of Operation.

When the team dated the corals from the pits, they discovered that staghorn coral had grown more or less continuously at Coral Gardens Reef since at least 1915. The coral persisted for more than a century. It may have been growing much longer—the pits didn't reach the base of the coral thicket, so we don't know when the staghorn coral first started to grow.[37]

"This does not mean that an El Niño, hurricane, and/or other stress events did not result in coral death," the team explains in their 2020 paper about the research, "only that there is no evidence of a massive die-off or collapse in the record retrieved from Coral Gardens Reef."[38] There could have been small gaps in the record of coral growth, but no major gaps.

While the species was dying in reefs throughout the Caribbean, this staghorn coral thicket persisted. The corals were growing when Model T cars were rolling off the production line at Ford. They were growing when agriculture achieved an industrial scale and chemical fertilizers started flowing into Caribbean waters. They were growing when white band disease started killing *Acropora* corals in the late 1970s, when urchins died in 1983 and 1984, and during Belize's first documented mass-bleaching event in 1998. At least 13 hurricanes and tropical storms made landfall in the area after 1930, but none of the storms caused the reef corals to die en masse.

∽

It's December 2022, two years after Lisa Greer and her team confirmed that Coral Gardens Reef has been a refugium for staghorn coral for over a century, and I'm in Belize to join them for fieldwork at the reef.

From my hotel room balcony, I can see the reef zones off the east coast of San Pedro, Belize. They are much like the classic pattern, with a white stripe in the tropical Rothko where dark, deep water meets the turquoise shallows—a line of breaking waves at a particularly robust reef crest. Below the line of waves is a more or less continuous line of elkhorn corals, a rarity now in the Caribbean. I squint at the turquoise stripe on the protected side of those waves, the backreef lagoon, searching for patches of reef. One of those patches must be Coral Gardens, where staghorn coral has quietly persisted even as the species died in other reefs, including those nearby.

The reef crest and backreef patches are all part of the Mesoamerican Barrier Reef, which stretches from Mexico to Honduras. Al Curran tells me that he thinks that this is the most unique barrier reef in the world as we later walk along the narrow beaches and seawalls of San Pedro's eastern shoreline. This large town or small city bustles with activity. We pass students walking to school, tourists exploring, vendors selling snacks and souvenirs. All the while we can hear the hum of scooters and golf carts zooming around town.

Unlike fossil reefs, their modern counterparts often have our coastal towns and cities as neighbors. What happens in reefs affects coastal communities and what happens in coastal communities affects reefs. Walking along the shore of the town, we see evidence of this influence in both directions.

What makes this reef unique, Al explains as we walk, is that it's such an effective barrier, which is of incredible value to San Pedro and other coastal communities like it. As I compare the crashing waves offshore at the reef crest with the gentle water lapping on the beach at my feet, it's obvious how the reef is protecting the town. Without it, even modest storms could cause widespread coastal flooding and erosion, and storms are getting stronger as climate warms. Even with the barrier reef, seawalls are becoming an increasingly common way to protect coastal development in San Pedro. New construction along the coast often has a seawall—a stout concrete wall parallel to shore to prevent erosion and flooding—while older buildings often don't. A newer seawall has been built on the ocean side of an old cemetery in San Pedro. It's particularly tall and blocks the ocean view, but I suppose its denizens are not there for the view.

Along the San Pedro coast, we see evidence of low-slung homes and small hotels being replaced by multistory condo buildings and fortified seawalls. In one spot, I find a new condo building with six bright white stucco stories facing the Caribbean. It dwarfs the two-story home on stilts next door. The colossal condos are so new that they do not yet have residents and their windows still have labels on the glass. The only activity at the place is the ceiling fans on porches that spin lazily in the wind like pinwheels.

Coastal development, now common around the Caribbean, often causes problems for coral reefs, particularly when it adds an influx of nutrients and pollutants to the shallow marine environment from stormwater runoff and inadequate wastewater treatment. Coastal development also often removes mangroves, forests of scrubby trees rooted in shallow coastal water that are a benefit to both nearby coral reefs and coastal towns.

Al points out a couple of mature mangrove trees between seawalls along the shore, evidence that there may have been mangroves all along this coast before the urban development of San Pedro. Between roots shaped like flying buttresses, mangroves create a shallow-water sanctuary for juvenile reef fish.[39] The roots also trap sediment and components of runoff from land that threaten to reduce water quality on coral reefs.[40] They are of value to coastal communities, too, because they buffer the coast from hurricane winds and storm surge.[41] And mangrove trees can help slow climate change; they pull more carbon dioxide out of the atmosphere than most forests on land.[42] We look across the boat channel at the wild part of the coast to the north, which is still thick with mangrove trees. San Pedro's coast may have once looked similar, protecting it from hurricanes and being a home for juvenile reef life. Coastal development and destruction of mangroves pose problems for reefs at a local scale. Climate change and the problems it creates affect reefs at a global scale. Problems at both scales are important to tackle.

Across an expanse of beach, we walk through tangles of seaweed that form a squishy brown blanket. The seaweed, called sargassum, is the most recent Anthropocene problem to wash ashore in the Caribbean. Sargassum accumulates naturally in the center of the North Atlantic, but since 2011, flotillas of it have been increasingly found in the Caribbean[43] thanks to rising nutrient levels, the warming climate, and possibly other factors.[44] Coastal residents and businesses in San Pedro that can afford it have the sargassum raked up and carted away every day, which leaves the sand of their beaches with the combed lines of a Zen garden. But the sargassum is not just a nuisance on Caribbean coasts. It's also a danger to coastal marine life in habitats like coral reefs, seagrass beds, and mangroves because the seaweed causes a drop in oxygen levels in the water, as well as other changes to water chemistry, and it blocks the sunlight.[45]

∼

I'm not in San Pedro only to explore changes in the coast and dodge piles of sargassum. I'm here to see the refugium at Coral Gardens Reef and meet the corals that survived while other Caribbean reefs faltered over the past several decades. Now that we've explored what's known about how reefs have persisted in the past, it's time to look at a modern reef that has survived amid numerous struggles over the past century, and to consider what it will take for reefs to persist into the future.

In addition to their research digging into the reef's subfossil past, Lisa Greer and her colleague Karl Wirth have traveled to Belize each year to collect data about the current health of Coral Gardens Reef. Over the years, more than four dozen undergraduate students have joined them to grow their research skills and diving abilities while studying the reef and monitoring its coral cover. I'm tagging along with Lisa, Karl, and Al and Jane Curran in 2022 for data collection at Coral Gardens.

As I was flying to San Pedro from Belize City, I realized that the idea of visiting a modern-day reef refugium after months of researching the mystery of ancient refugia filled me with the same giddy anticipation that my fellow passengers were expressing about their plans to relax in the sunshine with cocktails. The small plane was packed, but I was probably the only person on it who would be spending her time in Belize visiting corals that have been sheltering in place for decades. Visiting reef refugia may be a rather obscure activity appealing to researchers and science writers, but I think of it as a subcategory of an increasingly common type of Anthropocene tourism: seeking out places that have not yet been destroyed. People race to Montana to see Glacier National Park before the glaciers have melted entirely (it's nearly too late for that), and to California to see redwood trees before they burn in wildfires (which are becoming more frequent). We want to see what we might not ever see again, perhaps to shift our baselines, opening our eyes to what the world was like before, or perhaps to mourn the loss.

Each morning of field research, our reef-bound group meets boat captain Narciso Valdez on the dock and packs scuba and snorkel gear, air tanks, research equipment, and lunches into the boat before heading to Coral Gardens on the protected side of the reef crest, about 10 minutes from San Pedro.

I find that to jump off the boat at Coral Gardens is to dive into the past and see a time before staghorn coral died in the Caribbean. I jump in, look down through my mask, and feel my baseline shift to a much healthier concept of reefs.

I've seen small amounts of living staghorn coral before, branched colonies the size of tumbleweeds, but I've never seen it form a thicket so vast that it extends as far as I can see underwater. Most of the staghorn coral I've met was fossilized in outcrops or in the death assemblage of modern reefs.

In a fossil reef, staghorn coral will often look like a pile of firewood kindling, the bumpy corallites abraded off the sticks to various degrees. Sometimes those sticks are still in life position, jutting out of the outcrop at all angles, the space between them filled with sand and shells. In a death assemblage on a modern reef, sticks of staghorn coral skeletons rest in sandy piles between heads of coral, encrusted and bored by other animals.

But here they are alive, displaying all the warmer shades of a golden retriever thanks to zooxanthellae within their tissues. The branches are round in cross section, more or less the width of a flute, with a goosebump texture where polyps live. The tip of each branch narrows to a point like a deer's antler and is white rather than the yellow to orange brown of the rest of the coral. The tip is where new coral polyps are

Lisa Greer sets a transect line across Coral Gardens Reef, Belize, in December 2022.

added to the colony. Because staghorn coral grows so quickly, its newest polyps do not yet have zooxanthellae and are colorless. Much like the new white condos in coastal San Pedro, the polyps at coral tips without any resident algae have a vacant look.

The reef is shallow, only 10 to 20 feet (3–6 m) at its deepest parts, but Lisa, Karl, and I are scuba diving since we'll need to spend hours underwater to collect data. A swarm of colorful little fish dives down into the recesses of the largest staghorn coral thicket as we approach. Karl had told me that if you remain still next to the thicket, the fish will reemerge. I try it and it works. They start to appear out of the nooks in the coral, first just a few and then they seem to be everywhere, hovering just above the framework like a tossed handful of confetti. I kick my fins to get a little closer and they zoom back into the coral. The same stillness and patience required for bird-watching is needed for fish-watching—stillness and patience that I lack, especially underwater.

If you are a small fish, a thicket of staghorn coral is an ideal home. All reef corals provide habitat for animals like fish, urchins, crabs, snails, and shrimp, but staghorn coral is particularly well suited for creating habitat because of its endless branches. Some grow in meandering paths. Others are pencil straight. Some fork often and others don't, which creates a maze of negative space within the coral in which other life can shelter.[46]

Floating above the largest thicket of staghorn coral, I find it hard not to conclude that it is entirely alive, but it isn't. It's not uncommon to overestimate the proportion of live coral when eyeballing, Lisa told me as we prepared to jump into the water and collect data about the live coral cover. Looking closely at the surface of the thicket from above, I can see that some parts of the branches are alive and others are dead, their surface covered with splotches of macroalgae, coralline algae, sponges, and other life. In reality, very few reefs are entirely covered with live coral: before Caribbean reefs started their steep decline, they had 35–50 percent live coral cover according to estimates.[47] The team has analyzed the live coral cover at Coral Gardens at least once a year for more than a decade.

To quantify the percentage of live staghorn coral, Lisa kicks her way across the reef unfurling a long tape measure to mark a straight line, called a transect, along which data about the percentage of live corals will be collected. Then she manages a meter-wide square made of PVC pipe, reinforced with duct tape at the corners, moving it one meter at a time along the transect as Karl and I snap photos of the coral within each square meter. We repeat this process on several transect lines. Back on dry land, Lisa and her students will digitally trace the live coral in approximately 140 photos, calculating the amount living within each square meter.

While the reef's staghorn coral thrived for a century without interruption, these annual surveys of the live coral reveal that this is changing. Lisa and her team documented that the staghorn coral in this refugium faltered in 2016 and 2017, losing nearly half of its live cover. The largest thicket suffered the most loss. Living coral cover was over 50 percent in the largest staghorn thicket when the team first started monitoring.

By 2017, it had dropped to about 18 percent live coral,[48] an average amount for a Belize reef,[49] but a shockingly low amount for this unusually healthy reef.

Two instruments nestled within the reef, each about the size of a box of Tic Tacs, held the key to why coral cover dropped at that time. After we collect the photos of live coral along transects, I spot Lisa looking for the instruments. She's upside down and peering under the coral at the edge of the reef. Within each small plastic box attached to dead coral branches and lightly encrusted with limestone and algae, electronics have been recording water temperature every hour, day after day. Every six months one set of instruments goes off duty, its data are downloaded, and another set gets zip-tied to dead coral branches in the reef. The temperature data from these instruments stretch back to 2012, when the monitoring started.

Both the temperature data collected in the reef and sea surface temperature data collected by satellites show that 2016 was unusually warm. It wasn't only heat that was the problem, but also when that heat occurred, Lisa explains. Water heated up earlier in the year and stayed hot much longer than usual, subjecting the corals to more heat stress. In recent years, water has typically heated enough to stress corals by mid- to late July at Coral Gardens Reef. But in 2016 the reef was under heat stress by mid-May.[50] Worldwide, the peak heat in reefs was unusually high between 2014 and 2017 because of the combined effects of El Niño and climate change, which caused widespread coral bleaching. While El Niño is natural, global climate change is intensifying its impacts on temperature in tropical oceans including the Caribbean. The staghorn coral at Coral Gardens may have bleached, losing the symbiotic algae that give each polyp its warm color and its main source of food, but by the time the team surveyed the reef, they didn't see bleaching, just less living coral.[51]

The corals were dealt another blow in 2016 when, after more than two months of heat stress, Hurricane Earl made landfall in Belize on August 4.[52] In the wake of the storm, nearby San Pedro was strewn with fragments of roofing and lumber, splinters and nails from damaged buildings, fallen trees and electrical wires. Those seawalls along the coast were put to the test. In contrast, the team found few coral colonies broken or overturned by the storm at Coral Gardens.[53] However, the amount of live staghorn coral had dropped. Lisa believes that the storm may have added insult to injury—corals, weakened by heat stress, could not endure the hurricane. As we saw at Discovery Bay, Jamaica, and elsewhere in the Caribbean, too many disturbances can prove deadly for reef corals.

That decline in healthy staghorn coral at the reef in 2016 and 2017 was like nothing that's happened in at least a century, according to the record of past reef growth from the subfossil pits.[54] However, in the years since the decline, the team found that staghorn coral cover stabilized and eventually started to increase in some areas of the reef. In 2019, the largest staghorn thicket had jumped up to 25 percent live coral cover. By June 2022 it had over 40 percent live coral. Over several years, the staghorn coral bounced back.[55]

"It's under threat," Lisa Greer warned after our visit to the refugium in late 2022. "But it still is remaining somewhat resilient." We had no idea at the time, but we would soon learn the limits of this coral's resilience. The decline in staghorn coral that the researchers documented in 2016 and 2017 would turn out to be a harbinger of things to come.

In December 2023, just one year after I helped the team with fieldwork, I would receive an email from Lisa with the sad news. She and Karl were again at Coral Gardens for their usual fieldwork. When they dove into the water, they found no living staghorn coral. The yellow-orange thicket of living corals had become stark stony skeletons with branches covered in algae and other encrusting life. What had been a beacon of hope for staghorn coral survival had become entirely a death assemblage. Unlike the staghorn coral mass mortality in the 1980s caused by white band disease, and the earlier decline that was likely caused by poor water quality, the demise of Coral Gardens' staghorn coral was caused by extreme heat.

In the months since I had been bobbing above the healthy staghorn coral thicket, the shallow waters of the Caribbean and Eastern Tropical Pacific had grown extremely hot, heating up earlier in the spring of 2023 and remaining hot later into the fall.[56] Water was starting to warm as I was writing a first draft of this chapter. By the summer, the Caribbean was roasting with the most dramatic marine heat wave ever recorded. In one spot near Florida, the water was reportedly as warm as a hot tub. Researchers worried. Experts held press conferences and webinars. Corals bleached. The water got so hot that some corals died without even bleaching. NOAA's Coral Reef Watch program expanded its scale of bleaching warnings to add even more dire warnings for the unprecedented (until then) conditions. Deep into the winter, corals in some areas remained bleached and researchers watched closely for signs of recovery. But at Coral Gardens there was no hope of recovery; the slender branches of staghorn coral were dead. Six months after Lisa shared the sad news, I would join her, Karl, and about a dozen students at Coral Gardens and see the destruction for myself—the crumbling gray skeletons that were being perforated by boring sponges and covered with splotches of macroalgae—and I would feel like I was at an undersea funeral for someone who had died unexpectedly and far too young.

Much like the heat that caused some coral death at the reef six to seven years before, the 2023 heat was a result of the combined effect of El Niño and climate change.[57] In coastal Belize, sea surface temperatures were over 86°F (30°C) by late May, which is the typical threshold for coral bleaching, and they remained hot until late fall.[58] At Coral Gardens, the water temperature reached a high of 92°F (33°C) on October 7, 2023, according to the instruments nestled in the reef. It was too much stress, even for these resilient corals. They hit a breaking point. What had been a refugium for staghorn coral for over a century was gone. Corals that had managed to survive white band disease, declining water quality, overfishing, and other Caribbean threats were decimated by hot water.

On the flight to San Pedro, I had been thinking about myself as an Anthropocene tourist with an itch to see places before they're gone, but I had no idea how little time this coral had left. When I first splashed into a wonderfully healthy Coral Gardens Reef, I wondered whether it would be able to keep thriving as climate warmed. At the time, I thought there was reason for hope. The refugium had a history of making it through tough times, didn't it? But there are limits to resilience.

In many ways, climate change is a threat to reefs like no other, but perhaps there are other refugia out there, places somewhat protected from hot ocean water much like Coral Gardens was protected from twentieth-century perils, places where corals can lie low until we get climate warming under control.

~

While reefs are vulnerable to many problems, climate change is projected to be this century's greatest threat to reef health.[59] According to the Intergovernmental Panel on Climate Change (IPCC), the majority of coral reefs are very likely to decline this century as climate warms, although the number of reefs that are lost will depend on how much we can reduce emissions and limit warming.[60] Rising ocean temperatures are already increasing the threat of coral bleaching and are also linked to increases in the rates of death from coral disease. While refugia during the height of the last glacial period protected corals in smaller and steeper environments, and recent Caribbean refugia offered protection from local threats like water pollution and overfishing, future refugia will need to protect corals and other reef life from both global climate change and local threats.[61]

The idea that refugia allowed coral reef species to survive wild changes in the past has provided hope and fueled a flurry of research to find possible locations of reef refugia in the future.[62] There's a practical reason to search for future refugia: the information can help focus conservation efforts, protecting the reefs that are most likely to survive climate change from local threats like overfishing and pollution.[63] That's the aim of the 50 Reefs initiative, which links reefs that are likely to be refugia in the future with conservation efforts that can help protect them.[64] Once we have stabilized carbon emissions and climate, reef life in refugia could potentially repopulate degraded reefs. It's a Noah's ark strategy for biodiversity conservation and requires long-range planning. This, of course, will work only if we address climate change before the Earth heats to a level that refugia can't endure.[65]

One way of finding future refugia protected from heat is to locate reefs where temperatures haven't spiked much in recent years. Oceanographer Jennifer McWhorter led a search for refugia in the Great Barrier Reef by exploring our largest source of information about ocean temperature: sea surface temperature data collected by satellites. NOAA uses sea surface temperature data to calculate the heat stress that corals experience in reefs around the world by comparing the current sea surface temperature with the typical temperature in a location.[66]

Looking at two decades of NOAA heat stress data across the Great Barrier Reef, Jennifer and her colleagues found that not all locations are heating equally. Some places have experienced less heat stress, which Jennifer identifies as present-day reef refugia.[67] This doesn't mean these locations have perfect conditions for corals; they're still affected by hot water, just less so than other locations on the Great Barrier Reef.

Most of the search for reef refugia that will be sheltered from climate change is, at its core, a search for places where cool water can mix with overheated water, which helps corals avoid heat stress and bleaching. Jennifer and her team found that areas with less heat stress in the Great Barrier Reef had water that is more mixed by currents and tides than water in other areas of the reef. In the Caribbean, another search for refugia found sites where currents could keep reef waters cooler off the north coast of South America, along Mexico's Yucatán Peninsula, and in the passages between islands in the Lesser Antilles and Cuba.[68] In both the Great Barrier Reef and the Caribbean, the currents are created by winds. Winds can mix water and they also cause upwelling of colder water from the deep. The ocean typically has cold layers of water below and warmer layers of water above, warmed by sunlight and heat in the atmosphere. Upwelling occurs where winds move surface waters, which allows deep water to rise up. If a lot of water upwells, as it does off the west coast of South America, surface waters are too cold and nutrient filled for reefs to form. But a small amount of cool, upwelled water could help reef corals avoid heat stress and bleaching.[69]

To figure out whether the not-so-hot spots in the Great Barrier Reef will persist into the future, Jennifer used five climate models from the

IPCC *Sixth Assessment Report*—computer simulations of the planet that can be fast-forwarded to project how climate is likely to change over this century.[70] Models can be incredibly helpful, yet they can also be a bit detached from what's actually happening on coral reefs. "The time I have spent in the field doing research on coral reefs has given me more meaning and connection to the models I work with," Jennifer tells me when I note that learning about coral reefs with models seems rather different from learning about them by scuba diving.

According to the models, some areas of the Great Barrier Reef are likely to heat more slowly over this century. Jennifer and her colleagues identified these areas that are somewhat sheltered from heat as future reef refugia. But climate models focus on a large scale. They don't include details near coasts where corals live, so she needed to make some adjustments. Jennifer added necessary details to climate models—smaller-scale processes like winds, tides, and the shape of the seafloor along the Great Barrier Reef.[71]

She and her team found that some refugia in the Great Barrier Reef would persist if we are able to keep global average temperatures to only 1.5°C to 2.0°C above preindustrial levels, the target that the global community is attempting to meet by the end of this century.[72] However, with that much warming there will be more extreme heat events and those heat events will be even hotter than they are now, which will kill many corals. In scenarios with higher emissions and more warming, the projected outcome for reefs is much more grim. With over 3°C of warming, it's likely that all refugia in the reef would be lost, with corals meeting the same fate as those at Coral Gardens.[73]

To keep reef refugia livable, we need to limit the amount our planet warms. To limit warming, we need to reduce greenhouse gas emissions. "I personally feel like we've got to stop diagnosing the problem and just work on the solutions like getting our emissions down," says Jennifer.

It would be logical to assume that refugia at and near the equator would be the most likely to fail as the Earth warms, but Jennifer discovered that it's actually more complex. Refugia in northern parts of the Great Barrier Reef, which are closer to the equator, are likely to last longer than refugia in other areas because winds are projected to increase, which would mix water and could prevent it from heating as fast.[74] However, refugia will likely be lost in southern areas of the Great Barrier Reef, farther from the equator, because winds are projected to die down at the hottest times of year. Less wind means less mixing of hot and cool water.

For shallow reefs, the sea surface temperature data that Jennifer uses to find refugia are probably a good estimate of the water temperatures that corals are living with, but the temperature of the sea surface might not be as relevant for deeper reefs. The temperature could be much cooler near the seafloor where corals are living than at the sea surface in some cases, so Jennifer is working on methods of estimating the temperature at the seafloor based on the sea surface. Where the ocean is mixed by currents and tides, the temperature at the surface might be similar to the temperature at the seafloor, she says. But if the water isn't mixed, there may be layers of water at different temperatures, and there may be much cooler water at the bottom where corals live than at the surface.

Processes that cool water down on deeper reefs aren't easily detected at the ocean surface. For example, in May 2010 heat caused large-scale coral bleaching off the west coast of Thailand. Sea surface temperatures measured by satellites showed that heat stress was high, but in pockets of reefs about 50 feet (15 m) deep the heat wasn't as bad. Corals were somewhat protected. In these areas, researchers believe that cooler and warmer water were being mixed by internal waves.[75]

Internal waves aren't the type that crash on beaches and make boaters seasick. They aren't at the sea surface. Internal waves are within the ocean at the contact between layers of water, a colder and denser layer below and a warmer and less dense layer above. Tides often start water moving in internal waves where layers meet. As it sloshes around, the water runs into seafloor ridges and seamounts, which break up the internal waves into smaller waves. In places where they cause deeper, cooler water from the layer below to lap up into reefs, internal waves can help prevent heat stress and coral bleaching.

To test the effect of internal waves on the Thailand reefs during the 2010 heat stress event, researchers collected data on bleached corals at twelve sites: seven reefs affected by internal waves and five that were not. Corals were stressed throughout the area at that time, so there were bleached corals at all sites, but locations exposed to internal waves had more healthy corals and more corals that were only pale, having lost some, but not all, of their zooxanthellae. Locations that didn't have the cooling effects of internal waves had more bleached and recently dead corals.[76]

Temperature data collected by instruments tucked into the reef showed far less heat stress at sites with internal waves than reported by satellites measuring sea surface temperature. The reefs were not overheating nearly as much as the surface waters would suggest.[77]

Where else in the world could internal waves help create refugia? Internal waves aren't huge, so searching the tropics for them presents a logistical challenge. And the waves are underwater, so they can't be easily identified by satellites. To find reef refugia made by internal waves, oceanographer Scott Bachman turned to a computer model.

"I'm not an ecologist. I'm not a biologist. I never will be, but I know computing and I know math," says Scott. He applied these skills to find internal waves using a very detailed ocean simulation, collaborating with marine ecologist and geologist Joanie Kleypas.[78]

Their challenge was how to look globally for reef refugia when both reefs and internal waves are small compared with the ocean. "You can run global models and say what is the overall situation going to be for reefs all across the world, but, if the resolution of that model is, like, 100 kilometers, then you can't talk about individual reefs anymore and you can't talk about refugia," Scott explains.

To find refugia made by internal waves, they needed to use the world's highest-resolution ocean simulation, made using the MIT General Circulation Model.[79]

An ocean model uses hundreds of math equations to describe the physics of ocean processes like internal waves. All types of models, including ocean models, are used to create simulations of a hypothetical day, week, year, century, or more. The model simulation that Scott analyzed shows all levels in the ocean, worldwide, at each hour of each day over more than a year. It would be impossible to see the internal waves if the simulation had fewer times recorded. For example, if there was a data point from each day or week rather than each hour, it would appear as if there weren't any internal waves.

It would also be impossible to see the internal waves if the simulation didn't include high spatial resolution, meaning that it has more data points across an area than most models. A model run at low spatial resolution is like a photo without enough pixels—it's impossible to see the details. Scott once tested a model, running it at different spatial resolutions to figure out what works best to see internal waves. When you go from a resolution of two kilometers down to a resolution of 500 meters, internal waves can be seen everywhere, according to Scott. "It's really remarkable how much they appear on a small scale."

The model identified scattered refugia formed by internal waves in Southeast Asia, the Coral Triangle, the Galápagos, and the Eastern Tropical Pacific.[80] It identified very few refugia in the Caribbean or Atlantic. But Scott thinks there are probably more internal waves out there than we know about because even though the model simulation had a high resolution, had it been even higher, more waves would have been visible. However, because higher resolution requires more calculations and more computer power, there are limits to model resolution.

While most ideas about future reef refugia involve cool, deep water moving into the shallows, there's also an idea that shallow corals could move into the deep. This is called the deep reef refugia hypothesis.[81] Corals with symbiotic zooxanthellae can't live too deep since they depend on sunlight, but if they could live in slightly deeper water where sunlight still filters through, they might avoid hot surface waters that cause bleaching.[82] These refugia may not be an ideal environment for shallow coral species but could potentially be good enough to preserve biodiversity until shallow waters are hospitable again.[83]

For example, in 2023, the Mesoamerican Barrier Reef may not have been as hot at deeper depths as it was at the very shallow Coral Gardens Reef. During a search for staghorn coral survivors in northern Belize reefs, after witnessing the destruction at Coral Gardens, Lisa Greer and Karl Wirth found healthy staghorn coral in the forereef, 40 to 50 feet deep. The living staghorn coral wasn't huge, but it was healthy and filled with zooxanthellae.

Deep reefs can also be found much deeper than those reached by scuba divers, even over 500 feet (150 m) deep.[84] Full sunlight doesn't penetrate that far, yet some sunlight gets through in clear water, enough for corals that depend on photosynthesis. The partially lit world of these reefs is sometimes called the twilight zone, and it might also be a reef refugium.[85]

Less is known about reefs in the twilight zone than their shallower counterparts because they are harder to reach, far down the sides of limestone platforms, potentially a few hundred feet below a typical wall dive. In fact, these deepest of reefs, possible refugia as climate warms, might be in some of the same places where corals were sheltering in place when sea level was low during the last glacial period.

The deep reef refugia hypothesis has its critics.[86] It's unknown whether a deeper environment will protect corals. There's evidence that corals do bleach in deep reefs when surface waters warm, possibly because the colonies have a lower threshold for bleaching.[87] Also, it's likely that some deep water will be only a temporary refugium for corals. Jennifer McWhorter has found that with a 3°C (5.4°F) rise in global average temperature, areas of the Great Barrier Reef to a depth of 164 feet (50 m) would no longer be a safe temperature for corals in many places.[88]

There's evidence that as water heats, reefs in waters clouded by sediment and detritus from runoff may be somewhat protected from heat. The idea that murky waters, also called turbid waters, can become refugia as the climate warms seems rather counterintuitive since ideal conditions for corals include clear water. Sediment in the water prevents sunlight getting to corals for photosynthesis. Also, sediment settling down onto the coral polyps can be harmful—abrading the tissues and making the coral polyps expend energy to get sediment away from their mouths. But as climate has continued to warm and heat stress on reefs has become more common, scientists have observed that coral bleaching can be less common in turbid waters of nearshore reefs than it is in other locations, presumably because the cloudy water keeps corals in the shade. For example, during a recent heat event in Australia's Great Barrier Reef, *Acropora* corals experienced widespread bleaching in areas with clear water yet remained healthy in areas with clouds of sediment in the water.[89] It's estimated that 9–12 percent of the world's reefs are in turbid water that might be slightly cooler as climate warms.[90] However, it's not known whether all coral species can benefit from sediment-filled waters.[91] Some might not be able to survive in the murk.

The search for future reef refugia is a search for places that are less likely to have chronic heat stress, but that doesn't mean places identified as refugia are always safe. There is always a chance of acute spikes in temperature in our warming world. Just as there can be a heat wave in an Arctic climate and a rainstorm in a desert, there can be heat stress in an area of reef that is less prone to heat. Models can help quantify how the frequency of heat stress will change for reefs in the future, but they can't project whether an acute heat event will cause mass bleaching and kill corals. "One widespread warming event, or a succession of warming events, could have severe and irreversible consequences eliminating any concept of refugia," Jennifer and her colleagues warn in a 2022 paper.[92]

But if enough reef refugia persist as waters warm, coral species may survive, and those species could repopulate reefs once climate change has been solved in the future. That could give reefs a future in the long term, even if the short term is a mess. If species survive only in isolated refugia until we get climate change under control, then most reefs will be gone later this century along with the protection they provide to coasts during storms, the tourism money they generate, and the fisheries they support. These ramifications have motivated some innovative people to take matters into their own hands, building refugia to help more reefs survive. In the next chapter, we'll explore these human-created refugia and other actions people are taking to help corals persist into the future.

## CHAPTER 8

# DESIGNING REEFS THAT CAN SURVIVE US

Kirah Forman-Castillo points down at the shallow water next to the boat as the engine idles. Lisa Greer and I follow the line of her finger, looking for signs of the coral nursery below the choppy waves. Kirah, a Belizean marine scientist, helped build this fledgling reef restoration project at Hol Chan Marine Reserve in Belize when she was leading the reserve's science and conservation projects. Here, she and her colleagues are growing corals to restore a section of the barrier reef.

After we finish fieldwork at Coral Gardens in December 2022, we head about two miles northeast to this reef restoration site. It's not far from the reserve's namesake channel through the barrier reef (*hol chan* is Mayan for "little channel"), an area just behind the reef crest where most of the coral cover has been lost. Kirah and the Hol Chan team intend to change that. Somewhere in the light blue water off the boat's port side, the reserve staff have been adding small corals to a reef that could someday become a human-built refugium, a place where corals and other reef life will have a good shot at surviving this rocky time in the Anthropocene.

All I can see off the side of the boat is the usual patchwork of colors on the seafloor—yellow and light brown spots of living coral, dark brown seagrass, white sand, and the lumpy gray skeletons of dead coral. I finally make out a right angle and then another under the choppy waves—parts of a rectangular structure, a nursery, where fragments of coral are growing into new colonies. The plan is for the nursery-grown colonies to someday cover dead corals on the seafloor with new life.

Captain Narciso Valdez anchors the boat and Lisa Greer, Karl Wirth, and I pull on snorkel gear and follow Kirah into the water to see the project.

The coral nursery looks like a table with its legs on the seafloor. Its top, three to five feet from the sea surface, is made of rope strung across at roughly foot-wide intervals. Tucked between the ropes' twisted strands are branches of staghorn coral and prolifera, fragments of larger colonies that were harvested from Coral Gardens and other nearby reefs. Suspended on their ropes like tightrope walkers, these corals can grow up, down, and sideways, unencumbered by the seafloor or neighboring marine life.

Just before we jump into the water, Kirah mentions that a nurse shark regularly hangs out below the nursery table. "Security," she calls him. Bobbing in the waves, I peer below the table through my mask and am disappointed to find only sand and seagrass below it. He isn't on duty. Maybe this is his lunch break.

The coral fragments above the shark station have been growing in the nursery for about six months when we visit. Several months later, their branches will fill the ropes enough to look like garlands of tinsel. The garlands will be taken off the nursery and tacked into a nearby area of the seafloor atop dead coral skeletons. The hope is that the nursery-grown corals will keep growing within the reef, eventually covering their dead ancestors with a new layer of life. Perhaps someday a geologist or paleontologist will find an unconformity between the old and new, evidence of the missing time after the corals died and before the reef was intentionally restored.

On one side of the nursery, the tabletop is covered with a flat grid of wire instead of parallel ropes. Concrete pancakes on the grid serve as the foundations for fragments of coral species that prefer to encrust solid surfaces rather than grow on ropes. Kirah swims over to one end of the table to inspect the corals on the concrete pancakes.

Peering through her mask and holding on to the edge of the table to stay below the water surface, she focuses on an area with five brown

Kirah Forman-Castillo inspects corals growing in a nursery at Hol Chan Marine Reserve, Belize, in December 2022. Photo by Karl Wirth.

fragments of pillar coral. Large colonies of pillar coral have bulbous towers several feet high. They look fuzzy because the polyps' long tentacles extend into the water. But these tiny pillar corals are not much larger than bottlecaps. Pillar coral tends to be incredibly susceptible to stony coral tissue loss disease (SCTLD).[1] While it was never prolific in the Caribbean to begin with, pillar coral numbers have plummeted as the disease has spread. Growing fragments of it in the nursery is an attempt to help save the species from extinction by making several colonies out of the few that remain in northern Belize.

Kirah picks up one of the pillar coral pancakes and removes it from the table. Unfortunately, it's infected. Some of the polyps with long tentacles are gone. The coral's bare skeleton is exposed. Just as a toddler might pick up a cold from another kid at day care, a coral fragment on one of the concrete pancakes in the nursery may pick up SCTLD from a coral on a nearby pancake. By removing the pancake of pillar coral that shows signs of disease, Kirah is hoping to stop the spread to the other fragments. It's not the first pillar coral fragment she's had to remove because of disease. Chances are slim that any of these pillar corals will survive.

"We thought we'd try," Kirah says with a shrug. The fragments in the nursery are a last-ditch effort to save what they can.

It's tempting to call the nursery corals "babies" because they are little, just getting started, and because they are in a structure called a nursery. But they are not babies. In fact, these corals are likely much older than all of us who are snorkeling around their nursery table. Some may be hundreds of years old.

The polyps on the fragments of coral in the nursery may have been around for only a few weeks or years, but they are clones of corals that are quite old. The tiny colonies in the nursery are fragments of other colonies, so they are the same genet (i.e., they have the same genetics) and thus have the same age as the colony they came from. A survey of *Acropora* coral ages in northern Belize reefs (including Coral Gardens) found that staghorn coral genets were 62 to 409 years old, elkhorn coral genets were 187 to 561 years old, and the hybrid prolifera coral genets were 156 to 281 years old.[2] These corals are young compared to the oldest *Acropora* coral found in the Caribbean, which is over 5,000 years old.[3] (However, the survey of Belize corals was done in 2013 and 2014. It's unknown whether these elderly genets survived the extreme heat in 2023.)

Corals are fragmented intentionally to create new colonies for reef restoration. To fill this project's nursery table with coral, the Hol Chan staff snapped off finger-sized branches of established colonies elsewhere in the barrier reef, placed them gingerly in a bucket, and motored them to the nursery by boat.

Fragmentation happens naturally in the wild when storm waves break corals and the fragments found new colonies, adding more clones to both the original colony and the new outpost. This is how corals reproduce asexually. Corals reproduce sexually as well. Most species spawn, sending gametes into the water that form new genets when egg and sperm come together. Given the ages of *Acropora* genets in northern Belize and the lack of young genets, most of their reproduction, at least in recent decades to centuries, has been through asexual fragmentation rather than spawning.

Coral nurseries have become common worldwide in the past couple of decades, and they are particularly common in the Caribbean as a way to restore reefs degraded over the past several decades. There are a variety of nursery designs, most made of PVC pipe or metal. Some nurseries are shaped like spartan Christmas trees with coral fragments hung from PVC branches. Others are shaped like tables or A-frames on the shallow seafloor. As extreme heat has become more common, nursery designs have been modified with shade structures and ways to move them into deeper water to help protect their corals from bleaching. In some locations, nurseries are used as life support for coral fragments created accidentally when ships run aground. In other locations, like Hol Chan Marine Reserve, they are used to generate corals that can restore sections of reef that have lost coral cover.[4]

Hol Chan Marine Reserve's nursery was built and then seeded with its first set of coral fragments in late 2020 by Kirah and the Hol Chan team in collaboration with Fragments of Hope, a nonprofit organization based in southern Belize that is helping Belize's reef managers, conservation groups, and local residents learn coral reef restoration methods.[5] In mid-2022, the first batch of corals in Hol Chan's nursery had grown enough to be planted into the reef. A second batch of fragments was added to the nursery once the first batch was moved to its permanent home.

The seafloor next to Hol Chan's nursery is littered with dead coral skeletons—mainly the thick branches and trunks of elkhorn coral. Punctuating the skeletons are numerous sea fans and several large heads of boulder star coral, partly dead from SCTLD and currently undergoing treatments for their disease—an antibiotic cream applied by hand to the colonies. It's in this seascape that corals grown in the nursery are affixed to the reef.

As we snorkel over the site, Kirah points out several living staghorn coral colonies between the dead corals. These corals live here naturally, she says—a sign that this location may have livable conditions for more staghorn coral, which is one reason the site was selected.

Attached to the larger pieces in the coral death assemblage are living colonies of staghorn and prolifera corals, one to two feet across. They were planted into the reef about six months before our 2022 visit—the first cohort of corals grown in the nursery.

An old, dead coral skeleton makes the perfect place to attach a new coral since it is up above where sediment can encroach. Old, dead coral skeletons are often covered with macroalgae, which is scrubbed off the limestone before new nursery-grown coral is attached so that it doesn't compete for space with the coral.

Kirah points to several very small pieces of staghorn coral tucked into cracks in a dead coral skeleton. These staghorn coral fragments are much smaller than those growing in the nurseries, just little digits of coral. Kirah seems pleased with them. Once we poke our heads above the water, she explains the idea that fast-growing species like staghorn coral may grow just as well when fragments are planted directly into the reef instead of first spending time in a nursery. To try this, her team wedged small fragments that fell off other corals into cracks, hoping they would survive. And they have. Now, months later, we can see that the staghorn coral fragments are alive and cementing themselves into place.

Hol Chan has been training a squad of local residents to be reef first responders after hurricanes and tropical storms. One of their tasks will be to salvage coral fragments broken off the reef in storm waves. They will wedge the fragments into the reef, much like the little fingers of staghorn coral that Kirah pointed out, giving the pieces of coral a better chance of survival than if left on the seafloor.

While Narciso steers the boat back to San Pedro, Kirah tells me that the coral nursery we just saw is one of five in northern Belize. As the boat bumps along, she points north in the direction of the nursery at Mexico Rocks and another in Bacalar Chico, a national park that's an hour and a half to the north by boat. She points south in the direction of the two nurseries at Caye Caulker Marine Reserve, Hol Chan's neighbor.

These are early days of Hol Chan's reef restoration. Even though staghorn and prolifera coral grow quickly, the colonies added to the reef so far cover only a small area. But there's a way to see how a younger reef restoration might eventually look: visit the first reef restored by Fragments of Hope, which has had an additional decade of growth and demonstrates that it's possible to build a reef refugium.

Karl, Lisa, and I say goodbye to the rest of the group and take a small propeller plane about a hundred miles south to visit Belize's first restored reef, in the shallow water around Laughing Bird Caye due east of Placencia, Belize.

On the flight to Placencia, where we'll meet up with the Fragments of Hope team, I peruse maps of what we're flying over, realizing that Laughing Bird Caye is only about 12 miles (20 km) south of the reef at Channel Caye that we visited in chapter 4 where Bill Precht and Rich Aronson watched staghorn coral die decades ago,[6] and it's also not far from the reefs where Katie Cramer's team found evidence of the decline of grazing fish centuries ago,[7] as described in chapter 5. Knowing this history of loss makes it all the more shocking when we see the thriving restored reef at Laughing Bird Caye the following day.

In southern Belize, the barrier reef is much farther from the coast than it is in northern Belize. While standing on the eastern beach of Placencia's narrow peninsula, it's not possible to see the reef crest or the dark blue deep water off the shelf edge. The expansive backreef lagoon is home to seagrass beds, reefs, and little cays, many of which are more mangrove trees than dry land. The geography of the backreef in southern Belize is complex. It's not a uniform expanse of turquoise. There are deep areas as well as shallow patch reefs. Our destination, Laughing Bird Caye National Park, is part of a *faro*, a small atoll within the backreef, about a dozen miles east of Placencia.[8]

We head to the restored reef with four members of Fragments of Hope's Placencia-based staff including founder and director Lisa Carne, local board of directors chair Dale Godfrey, who's also our boat captain for the day, and Amir Neil and Victor Faux, who combine their work restoring reefs with tour guiding, fishing, and other jobs. They are part of the core team that maintains nurseries, outplants corals into restoration sites, and monitors restored reefs. They also conduct workshops to train others on reef restoration techniques. Fragments of Hope has trained about 90 Belizeans to restore reefs, including Kirah Forman-Castillo and other staff at Hol Chan Marine Reserve. As of 2022, the organization had grown over 170,000 corals from fragments in 28 nurseries throughout coastal Belize, including the Hol Chan Marine Reserve nursery that we visited, and they have outplanted corals into more than 17 reef sites. They have also grown a network of people throughout coastal Belize who work part time on reef restoration projects.

As Dale steers the boat toward Laughing Bird Caye, scattered palm trees at its south end come into view before the island itself, which is just a modest spit of sand. Rooted in the white sand, barely above sea level, the palms surround the visitor infrastructure for this island national park: a palm-thatched *palapa* shading picnic tables, a ranger station, and a restroom. The north end of the island, a tangle of mangrove trees, is closed to park visitors.

Dale beaches the boat near the palapa, alongside a couple of other visiting boats. We wade off the beach and within a few minutes are snorkeling over coral nursery tables. Beyond the tables are expansive clusters of extraordinarily healthy corals—staghorn, prolifera, and some elkhorn coral. Swimming over them, I can see evidence of the coral death assemblage on which the new corals were planted, mostly elkhorn and staghorn coral skeletons, remnants of a reef that died, likely of the same cocktail of stressors that bludgeoned most Caribbean reefs in the last half of the twentieth century, as we explored in chapter 5. Now most of these skeletons are covered with a thick blanket of living coral grown from fragments in nurseries and then outplanted.

"I haven't felt this hopeful in years," Karl exclaims once we pull our faces out of the water and away from the healthy coral below. I agree, and yet I am struck by the fact that several days before, I said something similar upon seeing Coral Gardens Reef.

Never having seen Caribbean reefs in their heyday, I had accepted not-so-healthy reefs as the norm. My baseline was shifted away from what the ecosystem's corals were like in the mid-twentieth century and before. But while we were photographing quadrats in the heart of the healthy staghorn thicket at Coral Gardens, I felt my reef baseline shift in a healthier and more hopeful direction. Now I feel my baseline shift again while snorkeling over expanses of restored reef at Laughing Bird Caye. It had not occurred to me that baselines don't have to shift toward death and decay. It's possible for them to shift in a positive direction, although that rarely happens amid Anthropocene destruction. Events like the 2023 demise of Coral Gardens are far more common than growing and thriving reefs.

We kick our fins and peer through our masks, circumnavigating the island that is more or less surrounded by restored reef. Over 10 years, Fragments of Hope has outplanted tens of thousands of fragments of *Acropora* corals near Laughing Bird Caye, and those corals have expanded in size. As I float above thousands of white-tipped staghorn coral branches, the reef's most prevalent coral, I start to notice subtle differences. Some staghorn colonies have branches poking every which way. Others have dense, upright branches growing side by side like people packed into an elevator. Some have thick branches. Others have thin branches. There are also more subtle differences in the shapes of the growing tips, the shapes of the corallite goosebumps, the style of branching, and other slight variations.

All species have variations. Humans, for example, have a wide range of shapes, sizes, and colors. Some of the differences are genetic. Some are related to both genetics and environment. Snorkeling above the corals, I recognize that I had not seen variation before in staghorn coral, or perhaps any coral species. It's rare to see variations in an endangered species because it's rare to see many individuals (or in this case, genets) in the same place at the same time. The possibility of losing an endangered species like these corals somewhat overshadows the loss of variation within the species, but variation between genets is important too. If the number of genets decreases, the chance of losing the species increases.

Genetic diversity was one of the goals for this reef restoration, Lisa Carne tells me once we are out of the water and on the island. "From the beginning, we took genetics into account," she says, adding that genetic diversity is needed for sexual reproduction, which is needed for long-term survival and to jump-start reef recovery.

Fragments of Hope sought out different genets of each type of *Acropora* as they grew coral fragments in nurseries and then planted them into the reef. At this point, the shallow water around Laughing Bird Caye is home to 17 genets of staghorn coral, 30 genets of elkhorn coral, and 3 genets of prolifera that we know of. There could be more if new

Six months of reef restoration
Hol Chan Marine Reserve, Belize

Over a decade of reef restoration
Laughing Bird Caye National Park, Belize

Examples of how reef restoration can change the amount of coral on the seafloor over time, both from December 2022.

genets have been made by corals that spawned, sending gametes into the water that met and formed coral larvae. To help ensure that spawning leads to baby corals, Fragments of Hope recently collaborated with scientists who gathered coral gametes for in vitro fertilization, a project that requires late-night gamete gathering while the corals are spawning and then matchmaking egg and sperm in dishes. The new corals start off as larvae that eventually settle down onto special plates before they are added to the reef.

Lisa Carne points out several small elkhorn corals that spawned recently. Each colony is a foot or two across and cemented to the skeletons of their dead ancestors off the west side of the island. They were planted four years before, strategically placed in proximity to increase the chances of creating coral larvae from the gametes during spawning. It's possible that this coral matchmaking has led to new elkhorn coral genets.

The reef at Laughing Bird Caye has been growing on its own in recent years. Fragments of Hope monitors the reef, which is approximately one hectare (2.5 acres or 10,000 square meters), with a drone, taking photos from above on calm days to document the amount of coral on the shallow seafloor and keep track of how the coral cover changes over time. The 2020 drone photo showed that corals had expanded to cover 7 percent more seafloor than the year before. In the 2021 drone photo, corals covered 14 percent more seafloor than in 2020. While most of the expansion is growth of existing colonies and new colonies that have formed through fragmentation, there has also been coral spawning, so some reef expansion may be new genets of coral. This growth trend may have been disrupted by the 2023 extreme heat, although it appears that the reef fared much better than others in Belize, possibly because of its genetic diversity, as we'll discover in the next chapter.

Restored coral can have a cascade of positive impacts. Marine scientist Nadia Bood, senior program officer at World Wildlife Fund Mesoamerica, which has supported Fragments of Hope since 2009, has witnessed how life on reefs has benefited from restoration over the years. "We have also seen that those restored sites have biodiversity coming in. You can see different species of fish, crustaceans in those sites. Even at the nursery tables, you can see different organisms recruiting to the tables and using underneath the table as habitat as well."

At this restored reef, fish and invertebrates aren't planted like the corals. They just show up. The reef at Laughing Bird Caye bustles with activity and sounds like the crackling of grazing fish and the snapping of shrimp claws. Schools of grazing fish flow over the corals, occasionally crowding together to nibble at a spot of algae. Victor points out a tiny juvenile spotted drum fish lurking in the recess under a coral. About the size of a coin, not including its long, ribbonlike caudal fin, it swims back and forth repeatedly, its black-and-white stripes a blur. After Lisa Carne points out a Maya hamlet—a small, endangered blue fish species that lives only in the shallow waters of coastal Belize—I start to see them everywhere in the reef, each calmly loitering solo by thickets of coral, its color the perfect complement to the golden branches.

While Nadia supports Belize reef restoration, she also hopes that there's less need for it in the future, except to recover from the impacts of disturbances. "Ultimately, I would like to see that we are effectively able to manage our reefs, doing restoration where possible, so that in the end, like maybe 15 years from now, we don't have to be worrying about restoring reefs because we are effectively managing them." Managing reefs provides protection from local stresses like pollution and overfishing. Belize reefs are protected in a system of seven managed areas called the Belize Barrier Reef Reserve System, which makes up 12 percent of the barrier reef area. The system was designated as a UNESCO World Heritage Site in 1996.[9]

After circumnavigating Laughing Bird Caye's seemingly endless healthy corals, I ask Lisa Carne whether the site is fully restored. How do you know when a reef restoration is done?

"Well, it's really hard to say," she says. "What are you restoring back to?" It's challenging to know whether a reef is fully restored or whether we just think it's complete because our baseline has shifted and we haven't seen reefs that are any healthier. We may be grading on a curve without realizing it, deeming a reef to be healthy when it is actually less healthy than reefs once were. Overall, she is satisfied with the genetic diversity at Laughing Bird Caye's reef, but she would like to see more elkhorn coral coverage. "Part of the problem is not having a structure to put them on," she notes, describing the areas of loose rubble, smaller pieces of the coral death assemblage rolling around on the seafloor, which are not a stable place for the next generation of coral to attach.

The fossil record can be a road map as we rebuild reefs, according to reef geologist and ecologist Lauren Toth, who studies the marginal reefs of Florida and the Eastern Tropical Pacific described in chapter 6. "I think it's really important to know what species were important historically and what species can really build these lasting structures," she explains. When Lauren talks with reef conservation groups working in the Caribbean, she pushes for restoration of elkhorn coral, boulder star coral, and brain coral. "Those are tough ones," she says.

Staghorn coral has often been a focus of Caribbean coral restoration projects because it grows so quickly and creates habitat for fish and other animals,[10] but according to Lauren, the species often isn't preserved as fossils and subfossils because it's so fragile. "It's not contributing to that long-term structure in the same way as some of the boulder corals and the elkhorn corals are," she says.

Slow-growing species, like brain and star corals, have been more of a challenge to grow from fragments for reef restoration. However, a new method called microfragmentation allows corals to grow much more rapidly from smaller pieces, which is promising to speed up the process of growing these corals.[11] The method was developed by David Vaughan, who formed the Plant a Million Corals Foundation after he retired from Florida's Mote Marine Lab. The microfragmentation method starts with a fragment of living coral that is cut into pieces as small as a centimeter wide, using a band saw with a diamond blade. Those pieces are grown in tanks on pancake-shaped surfaces. In about a year, the corals are large enough to outplant into a reef.

Unfortunately, in areas of the Caribbean affected by stony coral tissue loss disease (SCTLD), which targets these slow-growing corals, difficult decisions have had to be made in the past few years about whether trying to grow these species from fragments is worth the effort at this time.

Many have put their microfragmentation plans on hold until the disease is less of a threat. As Kirah Forman-Castillo found in the Hol Chan nursery, it's impossible to keep all coral fragments free of disease.

Planting coral fragments to restore reefs not only brings back ecological functions but also jump-starts the reef geologically, which is why Lauren Toth wants to see the corals that grow the most robust skeletons, and grow fast, included in restorations. In Florida, where reefs have shut down geologically, as we saw in chapter 6, Lauren and her collaborators have found that restoration could potentially bring back enough limestone production for reefs to keep up with sea level rise.[12] They looked at how the rates of limestone production have changed over time in Florida Keys reefs and projected into the future based on restoration goals developed by NOAA's "Mission: Iconic Reefs" project, an effort to scale up reef restoration on seven reefs in the Florida Keys National Marine Sanctuary. While rock cores can give a direct measurement of how much limestone Florida Keys reefs produced in the past, Lauren's team needed to use modeling to look at future limestone production. Given the amount of restoration planned, they found that within a couple of decades the restored reefs could be producing limestone at levels not seen for 7,000 years in the Keys and could keep up with sea level rise over this century.[13] We could go from watching Florida reefs drown to watching at least some of them flourish.

The first phase of the Mission: Iconic Reefs project aims to plant enough corals over seven to ten years to get the seven reefs up to 15 percent coral cover, with a focus on planting elkhorn and staghorn corals because they are fast growing and not affected by SCTLD. The effort will expand to increase the diversity of corals and animals that graze algae like urchins and crabs. During the second phase, they will expand the diversity of corals on each reef over a decade to increase coral cover to an average of 25 percent.[14]

Restoration of reefs in Florida is going beyond corals to ensure there are grazing animals in the ecosystem, which, as we learned during Caribbean reef decline, are essential for keeping macroalgae in check. For example, researchers are finding ways to restore populations of long-spined sea urchins (*Diadema antillarum*) along with the corals. The number of long-spined sea urchins in most Caribbean reefs has remained low since the 1983 to 1984 die-off. That event illustrated how these spiky invertebrates play an outsized role in reefs, mowing down slimy macroalgae, which opens up space for corals that build reef structure, as we explored in chapter 5. If the urchins can be restored and macroalgae levels are low, corals will have a better chance of expanding.

In a lab near Tampa Bay, marine scientist Joshua Patterson and his team have been testing and honing aquaculture methods to bring more baby long-spined sea urchins into the world. This is no easy feat given their complex month-long larval period, which includes a phase when each larva has two long, fragile spines that cannot break without killing the larva.[15] Yet they've had some success: as of early 2024, his team had raised 5,403 urchins. This is not nearly enough for Florida reefs, but it's a start. When released into a reef, the small spiky orbs, miniature versions of the adults, scuttle about the seafloor in search of shelter.

However, there are limits to how much restoration can save reefs in the Florida Keys, depending on how quickly we are able to tackle climate change. Reef restoration could keep these seven Florida reefs from drowning, but only if we reduce carbon dioxide emissions enough to slow the rate of warming. With too much heat, severe coral bleaching on the reefs will happen every year and the benefits of coral restoration will be lost.[16]

The extreme heat in Florida's reefs in 2023 tested the resilience of Mission: Iconic Reefs restoration projects and forced project organizers to search for creative ways to help keep their nurseries of coral fragments cool enough to survive. During the summer of 2023, about 4,000 coral fragments were relocated from their shallow marine nurseries to high-tech tubs on land. Under a giant tent, in rows of tubs, the coral fragments bided their time,[17] refugees from the reefs until temperatures cooled. The team at Reef Renewal USA, another Florida restoration project, moved their coral nurseries from overheating shallow water to offshore locations in water about 60 feet (18 m) deep, which was about 2°F (1.1°C) cooler.[18] While countless corals were lost, some were saved thanks to a remarkable number of people who contributed their time, boats, and other resources to help corals survive the heat.

~

In another marginal reef setting, the Eastern Tropical Pacific off the west coast of Central America, a nonprofit organization called Raising Coral is growing coral fragments in nurseries and then planting them into Golfo Dulce on Costa Rica's west coast.[19] Despite the subpar conditions of this environment, the corals added to nurseries and restoration sites are thriving.

"The corals want to grow," says Joanie Kleypas, Raising Coral's director. After a career as a scientist researching how the ocean and climate are changing and how those changes are harming marine life, Joanie started Raising Coral to help reefs recover from our ill effects. She describes reef restoration as helping Mother Nature do what she naturally does.

Raising Coral is also taking steps to make the environment more livable for corals by combining reef restoration with efforts to improve water quality in Golfo Dulce. Sediment gets into the gulf via runoff from land, clouding the water—a major problem for corals. Decades ago, forests surrounding Golfo Dulce were replaced with agriculture and housing for the workers—the same midcentury change in land use that degraded the reefs that Katie Cramer studied off the Caribbean coast of Panama, described in chapter 5. Today, climate change is making the problem of sediment runoff worse by causing extreme rainfall events, which carry even more sediment into the water. As Raising Coral grows coral fragments in nurseries and plants them into the reefs, they are also engaging local communities to find ways to decrease runoff. For example, reforestation can help keep sediment from eroding, and mangrove restoration can help trap sediment near shore. As Joanie puts it, reef restoration needs to start on land.

Unlike the barrier reef in Belize, the reefs in Golfo Dulce aren't large enough to protect a coast from storms. They are small and patchy. But they are incredibly valuable to animals that need reef habitat. Joanie has noticed that each coral added to the seafloor is a major boon for animals like fish, shrimp, and crabs that depend on them. "These other animals just treasure a coral," says Joanie. She's had fish watching her as she outplants nursery-grown branching corals in Golfo Dulce. The fish hover around during the installation and appear to also stick around long term, as if they've called "dibs" on the spot. As soon as the corals get large enough, crabs and shrimp move in too. The corals become much-needed habitat. Raising Coral has brought the region's primary reef-building coral, *Pocillopora*, back to the area. There are now thousands of colonies in Golfo Dulce.

I meet with Joanie while I'm questioning what makes a reef a reef. Learning about a multitude of reef restoration projects has me wondering whether restored reefs are "real" reefs. And for that matter, what is a real reef, anyway? Does it matter if a restored reef is not as the reef was decades or centuries ago if it saves species and expands habitat? As Lisa Carne noted, we often don't know what we are restoring back to because of shifting baselines.

Joanie suggests the term "rehabilitation" might be more accurate than "restoration" for these projects. To restore means to return something to the way it used to be. For example, to restore a painting is to make it like it was when new. Even if our efforts don't return reefs to the state they were in before the Anthropocene, Joanie reminds me that reef rehabilitation is still helping the ecosystem and its species persist.

Projects worldwide are taking similar actions to rehabilitate reefs. Many use nurseries underwater like Raising Coral and Fragments of Hope. Some reef restoration projects grow corals in tanks on land, and coral seeding programs grow coral larvae to be released into reefs. But restoring every reef is not possible. Critics argue that the area of reefs that can be restored through these projects is tiny compared to the area of reefs that need help.[20] But if these human-built refugia are anything like the natural refugia that helped nearly all coral species survive ice age sea level swings, they could be enough to ensure that coral species have a future in the long term even if many reefs are lost in the short term. It's possible that they can provide enough space for corals to survive until we get climate change and local environmental problems under control.

~~

Given the time and effort needed to restore and manage reefs, and the looming crisis of continued climate change, some scientists advocate for alternatives to traditional conservation including artificial reefs,[21] which theoretically could help reefs, sustain fisheries, and provide coastal protection.

The term "artificial reef" is used for a wide variety of structures on the seafloor, from concrete orbs to discarded ships and army tanks, statues, limestone bricks, and igloos made of concrete blocks. All artificial reefs include hard structures that encourage marine life to congregate, but they are not all effective at becoming actual coral reef ecosystems.

It's debatable whether many artificial reefs should be called "reefs" at all. When researching restored (or rehabilitated) coral reefs, I was questioning what makes a reef a reef, but at least those are built with corals, which seems like a minimal requirement for a coral reef. If you are looking for a coral reef ecosystem, artificial reefs may disappoint, particularly those for which marine life is an afterthought, but they can be helpful in other ways. Some are built to be fish habitats. Some are designed as destinations for scuba divers rather than an ecosystem.[22]

One common way to create an artificial reef is to add our debris to the seafloor. This happens accidently at the site of shipwrecks. Dive to an old shipwreck and you'll find the spot bustling with life even if the surrounding seafloor is quiet. Fish dart in and out of portholes. The hull is spotted with patches of colorful encrusters. More recently, ships have been intentionally sunk to make artificial reefs, and for some reason, the tools of war wind up on the seafloor as artificial reefs more often than other detritus. Florida may be the epicenter of sunken warships as reefs: the USS *Spiegel Grove* was sunk in 2002 six miles off Key Largo, the *Vandenberg* was sunk off Key West in 2009, and the USS *Oriskany* was sunk in 2006 off Pensacola. Because it had been an aircraft carrier, the *Oriskany* is known as the "great carrier reef."[23]

These ships can be effective fish habitats[24] and their hard surfaces can become home to encrusting organisms, but they may not be ideal for corals. The ships are often sunk in water that's deeper than most reefs, which limits the corals to those that can live with lower light levels. If they have antifouling paint, which is meant to slow or prevent growth of organisms on the ship's hull while it's in use, corals are unlikely to take up residence.[25]

When is disposing of unwanted items in the ocean pollution and when is it creative reuse? It does not seem sustainable to sink ships and other objects, especially when materials like steel could be recycled and reused if they were not sent to the seafloor. Additionally, while efforts are sometimes made to remove toxic materials before items are sunk, if they are not completely removed, fish may ingest toxic chemicals, and then people who eat the fish absorb the toxins too.[26]

Off the coast of Fort Lauderdale, Florida, over a million used tires were dumped into the ocean to create an artificial reef in the 1970s. At the time, this was thought to be an environmentally friendly plan and an innovative use for old tires. In retrospect, it was a terrible idea. The material that secured the tires together weathered away, leaving loose tires to be tossed about during storms, damaging the existing marine life and leaching chemicals. Because rubber is flexible, corals don't attach well to tires anyway. Recovering all the tires has proven to be much more difficult than sinking them. Contractors continue to pull them up each year, but as of 2019, about two-thirds of the tires still remained on the seafloor.[27]

Can structures added to the seafloor become a coral reef? Advocates of artificial reefs argue that it just takes time for species to take up residence, and that most artificial reefs are too new to be fully inhabited by reef life.[28] But we don't know whether they can become hubs of biodiversity like wild coral reefs.

～

To help attract corals to degraded reefs and artificial reefs, new technologies and methods are in development that aim to lure coral larvae, which could help increase biodiversity on an artificial or a restored reef.

One strategy uses advertising that coral larvae can appreciate much in the same way that developers building subdivisions will advertise a new community. A land developer might use billboards, brochures, and websites that picture utopia, or rather the idea of utopia because the neighborhood isn't built yet. Projects developing the reef equivalent of this advertising are hoping to entice coral larvae to settle down based on the idea of reef utopia: making a location seem like it's a thriving reef.[29]

How do you advertise to coral larvae? You use sound: it turns out coral larvae are influenced by audio advertising.

Coral larvae float in the ocean as plankton, carried by currents and eventually settling down and starting the rest of their lives cemented to the seafloor. But the place where they wind up is not random. There's evidence that larvae actively choose where they are going to take up residence, which is amazing considering that each larva is a mere cluster of cells, tiny and simple. They don't have ears but can sense sound and are more likely to settle down where the seafloor sounds like a healthy, bustling reef.[30]

Most of a reef's soundtrack comes from fish. Some are crunching as they eat. Others make sounds to establish their territory or even just talk to other fish. Soldierfish, bright red with dark eyes, make little grunting noises as they loiter under coral overhangs. Damselfish make clicking sounds while vigorously defending their spot in a reef. Parrotfish crunch limestone as they eat. Shrimp add to the cacophony with popping sounds as they snap their claws together. Not all reefs sound the same and reef sounds can change with the time of year and time of day, but overall, reefs sound lively.[31] And larvae want to be at the center of the action.

With underwater speakers on the seafloor, it's possible to make a ghost town of dead coral skeletons sound like the Times Square of reefs. Doctoral student Nadège Aoki and sensory biologist Aran Mooney are testing whether that could help lure coral larvae into areas that have lost coral cover.[32] They selected a couple of degraded reef sites in the US Virgin Islands, placing an underwater speaker at one site to blare healthy reef sounds. The other site, without an added soundtrack, is for comparison.[33] If more larvae settle down where they hear the sounds of a healthy reef, and results so far suggest they do, then projects that aim to create human-built reef refugia could use sound to entice corals to take up residence.

Coral larvae also need surfaces where they can attach and thrive, yet not all materials used for artificial reefs may be equally desirable to corals. Newly engineered materials have been developed specifically to encourage corals and other encrusters through the texture of the surface or its chemical makeup. If they work, these materials could make artificial reefs a better home for corals.

Some artificial reefs are engineered to suit corals with intentionally designed textures and shapes. An elaborate example is the artificial reefs created by Australian industrial designer and artist Alex Goad. His studio, Reef Design Lab, collaborates with reef researchers to design structures that suit the ecosystem, such as MARS (Modular Artificial Reef Structure).[34] Once the jack-shaped pieces of a MARS artificial reef are attached together, they form a three-dimensional lattice that looks like a very orderly version of the branching staghorn thickets at Coral Gardens Reef and Laughing Bird Caye in Belize. The cratered surfaces of each MARS piece are made with ceramic created using a 3D-printed mold. According to Alex, 3D printing is particularly good at imitating organic shapes and the rough surface of a reef, and the ceramic encourages corals to attach.[35]

In 2018, a MARS artificial reef was installed in the Maldives and planted with nursery-grown corals to test whether the engineered structure could be used in concert with reef restoration. The structure was built near a coral nursery at Summer Island where coral fragments are grown on ropes and perched atop sticks rooted in the seafloor. The latter look a bit like coral lollipops. After a year of nursery life, the corals are attached to MARS. Four years after the artificial reef structure was added to the seafloor, its once-white ceramic surfaces are purple, yellow, orange, and brown with encrusting corals, coralline algae, and other life. Fish large and small dart through the openings in the structure.[36]

While on display primarily underwater, Alex's designs have also been exhibited on dry land in museums. At the Museum of Modern Art in New York City (MoMA), a sculpture of MARS lattice was displayed in the *Broken Nature* exhibit (2020–2021), which, according to MoMA, highlighted works that show how design and architecture can help us repair our relationship to natural environments.[37]

To provide sites for corals to attach over a particularly large area that lost hard substrates, a restoration project in Indonesia peppered a two-hectare (5 acres or 20,200 square meters) area of seafloor with 11,000 structures designed to stabilize seafloor sediments and provide attachment sites for corals. The structures, called spiders (even though they have six legs, rather than eight, projecting from a central point), are about two feet (60 cm) wide and made of steel rebar dipped in fiberglass and then coated with sand to make the surface rough for corals.[38] Eighteen coral fragments were attached to each spider when they were installed on the seafloor. Two to three years later, at least 42 coral species were growing in the area and coral covered more than 60 percent of the spiders.[39]

Combining ecology and engineering at an even larger scale, a multidisciplinary team at the University of Miami is planning a new generation of artificial reefs, called hybrid reefs, a project funded by the US Department of Defense.[40] The project aims to make reefs more resilient to climate change while also including artificial structures that would break storm waves and provide structural support for reef life. In 2022 the team started a multiyear project to design and develop hybrid reefs in Florida, taking into account reef ecology, coral biology, engineering, materials, hurricanes, hydrodynamics, and other factors.[41] University of Miami researchers installed two types of hybrid reefs in the shallow waters off Miami Beach in 2023—one made of hexagonal concrete tubes and the other a hollow trapezoid with attached limestone boulders—to test whether these structures can successfully dampen waves and become home to corals.[42] With the shift from artificial reefs that dispose of materials like used tires to artificial reefs specifically designed for corals and other reef life, there's hope that these structures can support reef ecosystems.

To help ensure that reef refugia—whether natural or built—are able to help species persist into the future, we need to limit climate change by reducing emissions of carbon dioxide and other heat-trapping gases. The longer it takes to stop climate warming, the more reefs we'll lose. Even thriving reefs likely have limited time before they get too hot for corals to survive. For example, Coral Gardens Reef, which we visited in the previous chapter, had been thriving until it was toppled by heat.

Researchers are trying to find ways to help reefs avoid heat stress and endure the next several decades until we can get climate change under control. The strategies they are developing are not substitutes for cutting carbon dioxide emissions to tackle climate change. But they may be able to buy some time. One of these strategies looks, improbably, like emissions spewing from the stern of a boxy ferryboat. Or at least that's what a test of the technology looked like in March 2020 off the Australian coast.[43] The clouds of white coming from the ship weren't pollution. They were released from a device on deck that looks like a snowmaking cannon—a typical sight at midlatitude ski areas, but uncommon in the tropical Pacific. What appeared to be air pollution was actually a mist of seawater, which could potentially help prevent coral bleaching on the Great Barrier Reef, part of an effort to keep existing reefs livable for coral as climate warms. This program, funded by the Australian government and the Great Barrier Reef Foundation, is an attempt to engineer refugia to help coral reefs survive climate change, at least for a while.[44]

To describe how the mist of seawater might help prevent coral bleaching, we need to climb out of the ocean and into the atmosphere. Researchers hope that the salt in the mist will help make clouds more opaque, shading the reef and keeping water temperature from spiking during the hottest times of year. According to lead researcher Daniel Harrison, modeling data show that if thicker clouds can cut 6 percent of incoming sunlight, coral heat stress would drop by half.[45]

Some clouds naturally shade the planet, particularly those low in the atmosphere like puffy cumulus clouds. They reflect some incoming sunlight back out to space. Conversely, high clouds, like wispy cirrus clouds that form at the top of the troposphere (the layer of the atmosphere where weather happens), tend to let sunlight get to Earth and then prevent heat from leaving, which warms the planet.[46]

Daniel's team is trying to enhance the ability of low clouds to reflect a bit more sunlight away from the Great Barrier Reef during peak summer heat when the risk of bleaching is highest.[47] These clouds reflect more sunlight when they have more tiny water droplets, making them more opaque. The water droplets within clouds, so small that they float in the air, form as water condenses around particles like dust, pollen, smoke, and sea salt, which are called cloud condensation nuclei (CCN). In theory, adding more CCN to the air—as the team was attempting with the mist of seawater—could lead to more tiny water droplets within clouds, which would make the clouds more opaque and prevent some sunlight from getting through. This is called marine cloud brightening because the extra droplets make the clouds look brighter white.

Does it work? We don't know. At this point the technology is being tested to see whether it can create the nuclei needed to make cloud droplets, and to see where they go once released. The CCN produced by spraying saltwater into the air went upward toward the clouds, which is promising since that's where they would need to be to make more cloud droplets and extrabright clouds. But we don't yet know how clouds will respond, whether they will reflect more sunlight, or whether there will be unintended environmental impacts.[48] A team of experts that ranked actions to sustain coral reefs assessed marine cloud brightening as potentially moderately effective but ranked it lower than most other actions in part because of its possible adverse side effects.[49]

While cloud brightening aims to prevent bleaching, other strategies might help corals survive when they do bleach. One such strategy: we can feed the corals.

Bleached corals have expelled their symbiotic zooxanthellae, so none of their sustenance can come from the products of photosynthesis that they usually mooch from the symbionts. To recover from bleaching, the corals have to survive until the heat subsides and they can get some of the algae back. If they can change their diet, they are more likely to survive bleaching.

Unlike most animals, corals have two ways of getting food, as you may recall. They typically get the bulk of it from zooxanthellae. But corals are also predators. They may be tiny and stuck in place, but they can hunt.

Corals hunt with their tentacles. Some species have a circle of short tentacles around their mouth that looks like a baseball mitt with extra fingers. Others have long tentacles that flail about like those inflatable dancing tubes that lure customers into car dealerships. The polyps stretch their tentacles up into the water and catch zooplankton that floats by, delivering the food to the coral's mouth.

The amount of food that corals normally get from their zooxanthellae and the amount that they get from hunting prey varies, but the majority typically comes from photosynthesis by the zooxanthellae. When bleached, some corals can increase predation so much that they get all the nutrition they need. Others increase predation somewhat, but it's still not enough. A study of Hawaiian corals in shallow water found that nonbleached corals got 12–25 percent of the food they needed by hunting and bleached corals gathered 25–78 percent of the food they needed by hunting.[50]

If corals can modify their diet when bleached, we might be able to make reefs more livable for corals during the hottest times of year by delivering plankton to reefs. However, it would take time and funding to pepper heat-stressed reefs with plankton and monitor reefs closely enough to know when they might need more plankton. Plus, more research is needed to ensure that adding zooplankton to reefs won't cause additional problems for the ecosystem.[51]

Even with natural refugia, human-built refugia, and efforts to keep the ocean livable for corals, many reefs will be lost this century because of climate change and local stressors. It's likely that some coral species will not survive, but there is also evidence that at least some corals might have what it takes to persist in a warmer world.

## CHAPTER 9

# SURVIVAL OF THE HEAT TOLERANT

The coral nurseries in Golfo Dulce on the west coast of Costa Rica are filled with fragments of *Pocillopora*, also known as cauliflower coral (although it only vaguely resembles the vegetable). The growing corals dangle from nurseries shaped like Charlie Brown's Christmas tree, with sparse branches emanating from a central vertical pole. They punctuate parallel ropes in nurseries that look more like the setup at Hol Chan Marine Reserve in Belize, which we visited in the last chapter. Golfo Dulce is in the Eastern Tropical Pacific, a region known for its marginal reef conditions.[1] It's a challenge for reefs to form in this region because of temperature swings, acidic water, and rapid bioerosion.[2] Yet, as Joanie Kleypas and her Raising Coral team are finding, it appears that some genets of cauliflower coral are more up to the challenge of this environment than others. In March 2020, Joanie and Raising Coral project manager José Andrés Marin Moraga dove in Golfo Dulce to check on the nurseries. They had heard about a bleaching event in progress and wanted to know how the coral fragments in nurseries were faring in the heat. When they inspected the nurseries, they found that coral fragments along nursery branches and ropes were experiencing all different degrees of bleaching. Some were still dark with zooxanthellae. Some were as white as their namesake vegetable. And others were in between, having lost some but not all of their zooxanthellae. The same spike in water temperature caused a variety of responses in the cauliflower corals.

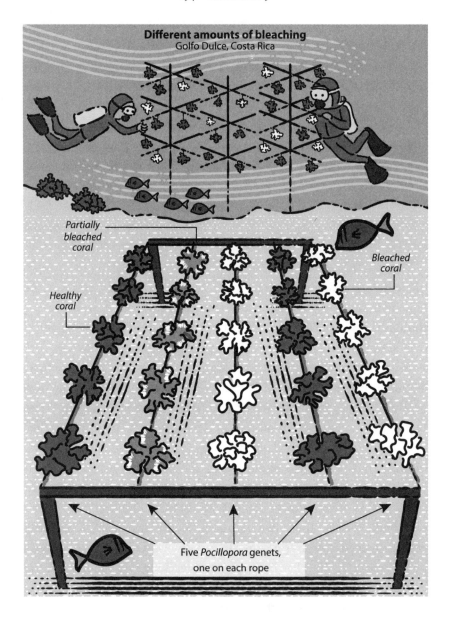

**Different amounts of bleaching**
Golfo Dulce, Costa Rica

Partially bleached coral

Bleached coral

Healthy coral

Five *Pocillopora* genets, one on each rope

There are a number of ways that corals, including *Pocillopora*, may be able to deal with heat, as we'll discover in this chapter. But the corals growing along ropes in Golfo Dulce illustrated why these particular corals had different responses to heat. Each rope of coral fragments held a single genet of *Pocillopora*. The corals along one rope were very pale, with little zooxanthellae left. The neighboring rope held corals that stayed a dark golden color, full of symbionts despite the hot water. Different genets had different reactions to the heat, Joanie and José Andrés noticed. Their colleague Tatiana Villalobos made a video of the rope nursery showing parallel coral garlands in a gradient of colors.

This happens in the wild, too—some corals bleach more easily than others—but in a controlled environment like a nursery with the coral colonies organized by their genetics, it becomes evident that some coral genets have what it takes to tolerate hot water and others don't. "It's so graphically visible," Joanie recounts.

The hot water caused an inadvertent test of which corals were the fittest and most able to survive. Corals adapted to live in warmer water are more likely to survive in a warmer world. This type of change—adaptation of species to tolerate heat—is not visible in the fossil record. The evidence of it has accumulated in the bodies of living corals.

Because reefs will have to contend with more heat stress in the future, reef restoration projects look for coral genets that have proven to be heat tolerant to make a restored reef as resilient as possible. Choosing corals that can persist in hot water and helping them reproduce is one example of a process called artificial selection, which humans have used for millennia to encourage species to evolve in more desirable directions.

Charles Darwin was fascinated by artificial selection, which is why he became enamored with pigeons. At the time, keeping pigeons, breeding them, and even racing them was a popular hobby in England. He was interested in how certain traits could be passed to offspring, so he got pigeons of his own, joined pigeon clubs, and peppered pigeon fanciers with questions about breeding.[3] Genes weren't yet understood, but the traits bred into pigeons were as visible as the nursery corals with different degrees of bleaching. Pigeon fanciers would breed the birds to select traits like the colors of feathers, the shape of the beak, and even elongated eyelids and large feet. Since their decisions about what traits to breed were artificial selection, Darwin called the process happening in the wild natural selection.[4] Today, natural selection is often known by the catchphrase "survival of the fittest."[5]

Through natural selection, a population of a species changes over many generations to better fit with its environment.[6] For this to happen, there must be variations in the population—different traits and genes—which can be passed along to future generations. Some traits are helpful while others are harmful (and some are neither helpful nor harmful). An individual with traits that are a good fit with the environment is more likely to survive long enough to reproduce, potentially passing those helpful traits to the next generation. An individual with a harmful trait is more likely to die young and be unable to pass the trait to the next generation. Over time, helpful traits become more common and harmful traits become less common.

While Charles Darwin usually gets credit for coming up with the theory of natural selection, another scientist, Alfred Russel Wallace, developed the same theory independently at almost exactly the same time from halfway around the world.[7] Wallace wrote up the idea as an essay while he was researching in Southeast Asia and sent it to Charles Darwin asking for feedback. Darwin was horrified when he opened the mail and found Wallace's essay because he had been trying for years to write a book about this same theory.[8] He had been thinking about evolution for two decades, ever since his voyage on the *Beagle*. There could have been a fierce rivalry between the two men with accusations about stolen ideas and arguments about credit, but in the end it appears that there wasn't.[9] Charles Lyell, the same geologist who was overjoyed when he realized that Darwin's theory of coral atolls would supplant his own (as we saw in chapter 2), seems to have helped Wallace and Darwin avoid ruffled feathers. He quickly arranged for Wallace's essay and some of Darwin's writing to be published together as one article on natural selection and wrote an introduction to the multiauthor article with botanist Joseph Hooker, describing Darwin and Wallace as "having, independently and unknown to one another, conceived the same very ingenious theory."[10] About a year and a half later, Charles Darwin's book *On the Origin of Species*, finally complete, came out and made a big splash.[11] While Darwin and Wallace did not always agree, they both explained and defended natural selection for the rest of their lives.

As the theory of natural selection spread throughout England, a now-classic example of natural selection was quietly occurring: the adaptation of peppered moths to environmental change. Before coal burning intensified in 1800s England, the tree trunks where the peppered moths perched during the day were light in color. The moths, their wings speckled light and dark, had more light spots than dark because darker moths, not well camouflaged against the pale bark, were more likely to be seen and eaten by predators. Once soot from burning coal darkened tree trunks, light-colored moths were no longer the fittest because they stood out on tree trunks and were more likely to be eaten. Peppered moths changed over many generations so that the majority had more dark spots than light. More recently, environmental regulations put in place in the mid-twentieth century have reduced the amount of soot. Tree trunks have lightened and the moths have become lighter in color as well.

With change in their environment, peppered moths adapted. Today, corals are faced with changing ocean temperature as climate warms, and adaptation could help them survive as well. Because natural selection happens over many generations, species like moths with short generation times (the time an animal lives before it reproduces) are able to adapt in decades or centuries. Species that reproduce very quickly, like bacteria, can adapt even faster (which is how some types of bacteria quickly become resistant to antibiotics). But animals like corals, with longer generation times, need more time for adaptation, and as climate continues to change, time is running out.[12] Because it's unknown whether corals have enough time to adapt naturally, artificial selection is being used to help speed the process as we will see. But we do know that, given enough time, corals have been able to adapt naturally to heat in the past.

～

It took thousands of years for corals in the Red Sea to adapt to heat through natural selection, according to scientists. The long, narrow Red Sea between northeast Africa and the Arabian Peninsula is home to 359 coral species and multitudes of fish that thrive in reefs fringing its coasts.[13] Despite the sea's extreme heat, the corals rarely (if ever) bleach, especially in the northern end of the Red Sea, which has earned them the moniker "super corals."[14] During the last glacial period, when sea level was low (about 20,000 years ago), the strait that today connects the sea with the Indian Ocean, the Bab el-Mandeb, was mostly above water, damming off the Red Sea. Isolated, much of the sea's water evaporated, which decimated its reefs. But around 8,000 years ago, once glaciers melted and sea level rose, Bab el-Mandeb flooded again and corals got back into the Red Sea from the Indian Ocean.[15]

It was as corals were traveling from the Indian Ocean into the Red Sea that they evolved to cope with high heat stress, according to research led by marine biologist Maoz Fine. He and his collaborators propose that corals passed from south to north through the hot, near-equatorial waters of the Gulf of Aden and into the southern Red Sea, with coral larvae reaching farther north over many generations.[16]

As corals moved from the Indian Ocean into the southern end of the Red Sea, where heat stress is extremely high, they adapted to high heat through natural selection—those least prone to heat stress were the most fit and most likely to survive.[17] Eventually, corals reached the Gulf of Aqaba at the northern end of the Red Sea where the water is not quite as hot. Because of the time they spent in the heat of the southern Red Sea, these corals were adapted for life in much hotter water. In the somewhat cooler waters of the northern Red Sea, their superpowers may have gone unnoticed were it not for Anthropocene climate change, which is warming the water. Reefs there are now faced with more heat than usual, yet the corals in the Gulf of Aqaba aren't prone to bleaching.[18] Adapted for hotter water, they have been able to persist, at least so far.

Coral tends to be at risk of bleaching when water is nearly 2°F (1°C) warmer than the typical maximum temperature in a place. In water that's too hot, zooxanthellae within the coral create toxins. The coral expels them to stay safe, bleaching in the heat. However, corals in the Gulf of Aqaba won't bleach until water is much hotter.[19]

To assess whether corals are likely to bleach in a particular place at a particular time, scientists calculate the degree heating weeks (DHW), a measure that considers both the amount of heat and the amount of time it has been hot over the previous three months (because heat stress builds over time). For example, if the temperature has been 1°C above

# SURVIVAL OF THE HEAT TOLERANT

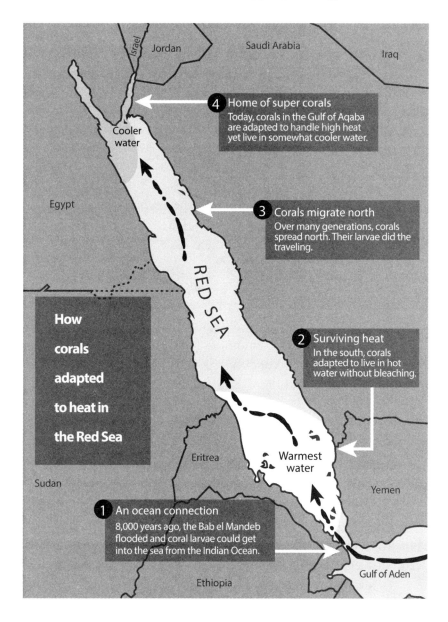

**How corals adapted to heat in the Red Sea**

1. **An ocean connection** — 8,000 years ago, the Bab el Mandeb flooded and coral larvae could get into the sea from the Indian Ocean.
2. **Surviving heat** — In the south, corals adapted to live in hot water without bleaching.
3. **Corals migrate north** — Over many generations, corals spread north. Their larvae did the traveling.
4. **Home of super corals** — Today, corals in the Gulf of Aqaba are adapted to handle high heat yet live in somewhat cooler water.

the typical maximum temperature for four weeks, DHW would be 4 (1°C × 4 weeks). If the temperature has been 2°C over the maximum for three weeks, DHW would be 6 (2°C × 3 weeks).

For most reefs, by the time DHW reaches 4, the more sensitive corals are likely to bleach. By the time DHW is 8, bleaching is widespread and corals are dying.[20] But corals in the northern Red Sea don't follow this pattern. Even when DHW is 18 to 24 (6°C hotter than average for 3 to 4 weeks), they do not bleach.[21]

To see what happens when corals from the Gulf of Aqaba *do* bleach, geoscientist Andréa Grottoli led a lab experiment that put coral fragments into hot water for over a month (the equivalent of 32 DHW). She and her collaborators found that the corals bleached in that heat, but they had adaptations that helped them survive. For instance, one species ate more zooplankton than other corals, so it was less affected by the loss of its symbionts, while another species had stored energy reserves that allowed it to persist while bleached. Different species had different adaptations that helped them cope, and overall the corals could maintain their energy.[22]

Because of these corals' heat-tolerance superpowers, Maoz Fine says the Gulf of Aqaba corals may be the last corals to survive in the heat.[23] He calls the reefs there a refugium.[24] Unlike refugia defined by their livable environmental conditions (chapter 7), this refugium's defining characteristic is its coral superpowers. "These reefs offer such a great hope and a stockpile of immense biodiversity," Maoz said in a 2022 presentation.[25] Depending on when we get climate change under control, the waters of the Gulf of Aqaba may eventually warm too much even for these super corals, but it's also possible that some corals will survive.

A major ingredient needed for natural selection is time—lots of it. Natural selection would be impossible without the ability of small changes to add up over a long time. Corals in the Red Sea adapted to handle heat over thousands of years, but climate is warming much more rapidly now because of the excess greenhouse gases we've added to the atmosphere. This may give rapidly reproducing creatures like bacteria and moths time to evolve, but not corals.

Given the projected rise in temperature this century, corals have only a limited time to adapt to heat naturally—decades instead of millennia—but there's evidence that human assistance could help speed up the process. Human-assisted evolution uses artificial selection to intentionally breed certain characteristics into populations—for example, breeding corals for heat tolerance.[26]

Many genes appear to have some bearing on a coral's ability to survive heat.[27] Increasing these genes in coral populations could help as climate warms. Human-assisted evolution has typically been used to breed domesticated animals—dogs, farm animals, and Darwin's pigeons—with certain characteristics, so some experts caution that more research is needed to understand the possible negative consequences of assisting coral evolution in the wild.[28]

Human-assisted evolution can mean breeding certain coral genets in laboratories and then releasing them into the wild. For example, one project enhanced the heat tolerance of Indian Ocean brain corals by breeding them in a lab with brain corals from the Persian Gulf, which are adapted to live with high heat. The offspring have more heat tolerance on average than their Indian Ocean parents.[29]

But laboratory conditions are not always necessary to enhance coral heat tolerance. Reef restoration projects often combine local coral genets with genets from sites that have more challenging environmental conditions in order to increase variation in the coral population and potentially infuse it with helpful genes that could create more resilient coral genets.[30]

For the reefs that Fragments of Hope is restoring in Belize, like the reef at Laughing Bird Caye, Lisa Carne and her team have added corals with a demonstrated ability to cope with spikes in temperature. The traits that allowed corals to survive past heat can, they hope, better equip the coral population to survive future heat.

After our snorkel around the restored reef at Laughing Bird Caye, the Fragments of Hope team takes Lisa Greer, Karl Wirth, and me to a spot where they harvested fragments of corals with heat tolerance—a shallow area of the backreef near Placencia where water regularly warms to bleaching levels and yet corals have persisted. In our snorkel gear, we hop overboard into calm, shallow water next to a small anonymous clump of mangrove to see the reef. Right below the surface are thriving

*Acropora prolifera* corals, numerous fish and urchins, and the crackling noise of a lively reef.

Fragments of Hope learned about many similar patches of coral from local fishers, including the organization's staff, who know what's on the shallow seafloor and where corals rarely bleach. Harvesting corals that can survive in heat, and using those corals for restored reefs, they hope that heat tolerance traits will be selected as the corals adapt, making the reef more resilient as climate continues to warm.

This strategy was put to the test as intense heat enveloped reefs in 2023. The water remained hot for months. Unlike the Coral Gardens Reef, a hundred miles to the north, where the staghorn coral was killed in this event, many of the reefs restored with heat-tolerant corals showed some resilience. Once the heat started to subside, the Fragments of Hope team surveyed their restored reefs and found that while quite a few corals didn't survive, many held on to their warm orange and yellow colors despite the heat, or bleached and then recovered.[31] In photos from their surveys, colonies that bleached completely, turning ghostly white, sat next to colonies that remained healthy.[32] They shared the same hot water but didn't have the same ability to cope. There are winners and there are losers, Lisa Carne told a network of Caribbean reef experts as she shared photos of healthy and bleached corals.[33] As heat subsided in late 2023, the Fragments of Hope team was figuring out which genets were the survivors. Their future reef restoration will rely on these fittest corals. Their artificial selection work is ongoing.

For natural selection to occur in the restored reefs, the corals must sexually reproduce. If the coral genets that her team brought together at Laughing Bird Caye breed, then they could adapt over time. This is why Lisa Carne thinks about coral sex more than most people. "How are you going to adapt if you don't have sex?" Lisa Carne says with exasperation at the idea espoused by some that asexual reproduction of coral clones is all that's needed to restore a reef—growing new colonies from fragments of existing colonies. If corals reproduced only asexually, there would be no way for novel combinations of traits to pop up. "You cannot have evolution and adaptation if you're not crossing your chromosomes with something else!" she proclaims.

In the Caribbean, with multiple threats to coral health, heat tolerance isn't the only helpful trait. High growth rate, low susceptibility to disease, and a high rate of sexual reproduction are beneficial in a variety of conditions.[34] Corals that have survived multiple stressors over the past century might have adaptations that could help them survive into the future, especially if they also have heat tolerance. Using these corals for restoration is hopefully spreading resilience into more reefs.[35]

After we clamber out of the water onto Laughing Bird Caye, I pepper Lisa Carne with questions at the water's edge about her coral matchmaking efforts. While she isn't optimistic about much, she is hopeful that the corals outplanted into restored reefs are spawning and creating new genets that could have what it takes to survive. She looks out at the shallow water around the island where we sit and wonders aloud what new resilient corals might be out there.

Will corals be able to adapt to anthropogenic climate change over this century? We don't know for sure. A third of coral species are threatened or endangered today, with dwindling populations.[36] It's unlikely that all coral species will make it through the increasing stress we're creating as we pour carbon dioxide into the atmosphere, making the ocean warmer and more acidic, on top of local strains like pollution and overfishing. But it's also possible that at least some corals have the helpful traits they'll need to survive the heat.[37] We just don't know whether there will be enough time for these beneficial traits to become common in coral populations.

However, corals have some faster ways to improve their heat tolerance. In addition to their ability to adapt over many generations, they also have ways to become more resilient within their lifespans because the environment they live in molds them as well as their genes.

An animal's DNA and genes typically stay the same throughout its life, but its gene expression can change. If an animal's genes are like sheet music on a piano stand, its gene expression is the notes the pianist actually plays at any given moment. The pianist may not play some notes, just as an animal may have genes that are unexpressed.

Our gene expression can change in response to the environment we live in. For example, when people live at high altitude where air is thinner, their bodies become more efficient at collecting and using oxygen. Another, albeit rather unusual example: twin astronauts Mark and Scott Kelly looked far less identical after Scott spent a year in space—a very different environment. NASA estimates that Scott's time in space caused a 7 percent change in his gene expression.[38]

The ability of an individual to adjust gene expression to cope with environmental conditions is called *acclimatization*. Unlike adaptation, in which a population changes as genes and traits are inherited over many generations, acclimatization allows an individual to adjust its ability to respond to the environment during its lifespan. Not all living things can acclimatize, but there's evidence that corals, at least some, have a remarkable capacity for their gene expression to be shaped by the environment they live in.[39]

There's also evidence that changes in gene expression that make a coral more able to cope with heat may be inherited by the next generation of corals.[40] This inheritance of gene expression, known as epigenetic inheritance, may allow corals that are acclimatized to hot water to pass along a way to cope with heat to their offspring, which essentially inherit resilience. Epigenetic inheritance also affects other organisms, including humans, and may cause all manner of things to be passed down, including inherited trauma. While there's a lot we don't yet know about how trauma might be passed to offspring, it's possible that later generations aren't necessarily afraid of what traumatized their ancestors. They may instead have a heightened ability to sense it, which could be helpful in avoiding harm in the future.

Embedded within the idea of natural selection is the idea that each creature is an individual. Read a description of how natural selection works and you'll find the word "individual" used over and over. Individuals are within a population. Individuals are different from each other. Individuals may have traits that are beneficial. Individuals may have traits that are harmful. Individuals may be more fit than others and more likely to survive. Individuals that survive reproduce, making more individuals. Survival of the fittest implies that each individual is out for itself in a competitive world. But for some species, it's hard to define what even constitutes an individual.

Corals are not individuals. They are collectives. In a colonial coral, there are individual polyps, each living in its own limestone penthouse, but those polyps work together. They help each other out. Polyps are connected by tissue bridges that they use to share resources like food. Beyond individual polyps, there are other members of the collective within the coral tissues like the photosynthetic zooxanthellae as well as a complex suite of bacteria, archaea, fungi, and viruses.[41] Together, all this other life is a coral's microbiome, and the makeup of the microbiome may have an important impact on a colony's ability to survive in heat.[42]

Corals aren't unique in sheltering microbes within their bodies. Many animals have microbiomes, although some have only a small collection of microbes. Humans have large microbiomes, with 10 trillion to 100 trillion tiny symbiotic organisms within each person.[43] But coral microbiomes are even larger. For example, there are over 10 times as many microbes on the surface of a coral as on a similar-sized surface of human skin,[44] and studies suggest that the diversity of coral microbiomes is higher too.[45]

The coral colony with its interdependent polyps, its photosynthesizing zooxanthellae, and its large and diverse microbiome add up to a system instead of an individual. Because of the complexity of this system, coral collectives have additional tricks that could help them survive in our warming world beyond evolution by natural selection.

Exactly how a microbiome supports a coral colony's health isn't yet well understood, but we know that some microbes within coral actively help their host, and multiple microbes appear to help protect corals against heat stress.[46] Other microbes just take up space, which may be helpful if, by taking up space, they prevent harmful microbes from getting into the coral.[47] In the Caribbean, researchers have found that elkhorn and staghorn corals were less likely to be affected by disease when they had high levels of a certain bacterium in their microbiome.[48]

Corals' microbiomes can be disrupted for a variety of reasons, and once these are disrupted, corals become more vulnerable to disease. It's possible that the outbreak of white band disease that killed vast amounts of staghorn and elkhorn coral in the Caribbean in the late 1970s and 1980s was made possible by a shift in coral microbiomes related to earlier stressors, such as an influx of nutrients from fertilizers, which started the decline of these corals in the 1950s and 1960s, as we saw in chapter 5.

Heat stress can also cause the microbiome of some corals to shift to include less beneficial microbes and more pathogens.[49] Losing part of the microbiome can be like a key member of the team quitting. It's possible that this role can't easily be filled by either the coral or another microbe, and the colony's health suffers.[50] Corals with some heat resilience seem to have stable microbiomes, even during heat.[51] Corals that can keep their microbiomes stable may be more likely to persist as climate changes, which may shape coral species diversity in the future.[52]

Because a coral's microbiome is important for its health, it's possible that dosing corals with probiotics could keep them healthy during heat stress and other environmental changes.[53] This could potentially make corals less susceptible to bleaching and disease; however, some argue that more research is needed to better understand which microbes are beneficial, which are harmful, and whether giving corals a dose of probiotics is even feasible at the scale that would protect reefs.[54]

The most high-profile member of the coral microbiome is zooxanthellae, the symbiotic algae that, through photosynthesis, help corals get the nutrition they need. Some researchers don't even include zooxanthellae with the rest of the microbiome because of their special role, but—microscopic and living within coral—they fit the definition.

The zooxanthellae in coral are unicellular algae in the genus *Symbiodinium* (and some in genus *Durusdinium*). There are many different types of zooxanthellae, and some can take the heat more than others.[55] The most resilient varieties often stay put in their coral during heat stress.

Other, less resilient types start making toxins when they overheat. The coral then kicks out the zooxanthellae and bleaches in the heat.

There's evidence that at least some coral species are able to swap out their zooxanthellae for a type that's more heat tolerant if needed. Lab experiments that send corals into hot water and watch what happens to the zooxanthellae have found that after a coral with sensitive symbionts bleaches, it may later take in mostly heat-tolerant symbionts, which can help the colony avoid bleaching again.[56] However, there is a trade-off—coral with heat-tolerant symbionts may grow at a slower pace, which could help the coral survive extreme heat events but could make it more difficult for reef building to keep up with sea level rise over the long term.[57]

This symbiont switching is known to happen in the wild too, which might help corals acclimatize to heat. Off the west coast of Panama, in the Eastern Tropical Pacific, scientists found that colonies of *Pocillopora* have either heat-tolerant zooxanthellae or a variety that isn't as heat tolerant. They tracked both symbionts and coral colonies in 2015 and 2016 during El Niño heat amplified by climate change and found that the colonies with more heat-tolerant symbionts were much more likely to survive.[58] Colonies with less resilient symbionts were more likely to bleach and die, but of the survivors, many swapped their symbionts for the more heat-tolerant type.[59] This shows evidence of adaptation by natural selection—the fittest corals survived, or actually, the corals with the fittest zooxanthellae survived. It also shows that some corals increased their resilience by procuring more heat-tolerant symbionts. The corals that bounced back with new symbionts should now be more fit and less likely to bleach in the next heat event.

∼

If you were to stand on the white sand beach on the south coast of Ofu, a remote South Pacific island in American Samoa, you would see mottled patches of corals and other reef life strewn throughout the placid shallow water before you in the backreef. Farther out, you would see waves crash at the reef crest. In this particular location, researchers have found evidence that a specific set of environmental conditions have helped corals both adapt and acclimatize to cope with heat.

Some backreef lagoons are deep enough to scuba dive, but this area of Ofu's backreef is very shallow—in some locations, corals are in less than two feet of water. Some places in the backreef are so protected by the reef crest that the water gets trapped at low tide, unable to mix with cooler water. When roasted in the sunshine, the shallow water heats up by 11°F (6°C).[60] But then the tide rises, cooler water flows in, and the temperature drops back down. The dozens of coral species living in the backreef endure seawater temperature rising and falling,[61] warming and cooling, day after day with the tides.

The temperature isn't uniform across the backreef. There are a number of pools in the backreef, a mile across and deep enough to swim, and some have more mixing of water, which keeps temperatures more stable during tidal cycles. In other pools with less mixing, the temperature swings wildly up and down.[62]

The first time that marine biologist Steve Palumbi saw the backreef pools, he realized that these reefs are too warm for corals to survive, and yet they do. Despite the tough conditions, reefs are thriving. Steve calls one of the pools "the village of elders" because the corals are all hundreds of years old and enormous. "They're just stunning," recalls Steve fondly, "because you know that pool has been a place coral can live every day of the last 500 years." Chances are, below the living elders are the skeletons of their ancestors, a record of corals that lived over millennia, a long lineage of elders.

For more than a dozen years, Steve and his collaborators have explored the questions of how and why corals can thrive in this environment despite the temperature swings. They've found that not only do the corals survive daily swings in temperature in the Ofu backreef, but there's evidence that these tough conditions have been like a coral boot camp, helping colonies develop the ability to deal with heat stress.[63]

To investigate this coral boot camp, Steve's team focused on the tabletop coral *Acropora hyacinthus*,[64] which can form colonies in the backreef with hundreds of short, chunky fingers extending upward from a broad, flat surface like little shag carpets. Common throughout the Indian and

Pacific Oceans, this coral is typically sensitive to environmental stress and vulnerable to bleaching,[65] but in Ofu's shallow backreef, it's surprisingly resilient.

Several years ago, Steve Palumbi and Thomas Oliver, a PhD student at the time, figured out that the swings in temperature with the tides help corals build resilience to heat stress. They compared tabletop corals that live in two areas: a backreef pool with a big daily range in temperatures and a pool with more stable temperatures. They put coral fragments from both sites into 50-gallon tubs with water hot enough to simulate the heat stress that corals will regularly need to endure as climate changes.[66] They didn't have a lab in Ofu, so Tom made a makeshift lab on the beach and monitored the corals in tubs from there. The coral fragments that had lived in a pool with wild temperature swings fared well in the hot tub— most survived, unlike the coral fragments that had lived in a pool with more stable temperatures, nearly half of which died in the heat.[67]

Did the corals from the site with temperature swings adapt through natural selection over many generations? Or did they acclimatize, becoming used to changes in temperature? To figure this out, Steve and several collaborators started a coral exchange program. Coral fragments from the backreef pool with more stable temperatures were moved to the backreef pool with wild temperature swings, and vice versa.[68] During the two years corals spent in their new location, the researchers tested how adept they were at resisting heat stress. They periodically put the corals into tanks of hot water as before (although by this time they had an indoor space for the experiment and no longer had to use a beach lab). Corals from the more stable pool increased their ability to deal with heat as they lived in the pool with large temperature swings, turning on genes involved with heat tolerance and demonstrating that some of the corals' abilities to handle heat came from acclimatization.[69] However, they didn't become as adept at coping with heat as the corals from the location with temperature swings—evidence of adaptation over time as well.[70] According to Steve, adaptation and acclimatization each account for about half of the corals' heat tolerance.

Looking at genetics, the researchers found changes in gene expression[71] (indicating acclimatization), as well as genes involved with heat tolerance[72] (indicating adaptation by natural selection). Over less than two years, acclimatization helped these corals develop as much resilience to heat stress as would happen by natural selection over many generations.[73] Corals from the two locations also had differences in microbiomes, including heat-tolerant zooxanthellae.[74] They grew their ability to resist bleaching, which is a hopeful sign that at least some corals might be able to survive long enough for us to get climate change under control.

There's a maxim that what doesn't kill you makes you stronger. I'm not sure that's necessarily true when it comes to humans, but it does seem to be the case for corals living in environments with regular temperature swings. Each heat spike at low tide in the Ofu backreef was well over the amount that puts most corals at risk of bleaching, but it stayed hot for only a few hours, so stress didn't accumulate. In their book *The Extreme Life of the Sea*, Steve and Anthony Palumbi describe this ability of Ofu corals to get used to heat with an analogy: "In the classic film *The Princess Bride*, the hero has trained himself to resist poison in a similar way: gradual exposure, day by day, until once-lethal doses grow laughable."[75]

It's not known whether all coral species can acclimatize or adapt like the tabletop coral, but given that other coral species live in the Ofu backreef boot camp, there are likely more species with these abilities. And the Ofu reefs are not unique. Coral bootcamps with daily swings in temperature have been found throughout the tropics,[76] which may be helping corals build resilience to heat.[77]

Corals that live with daily swings in temperature, in Ofu and elsewhere, may have resilience to heat, but because the locations are usually in shallow water near shore, the reefs are vulnerable to other threats. Near developed coasts, corals may have to deal with runoff that brings sediment, sewage, fertilizer, and other pollutants into the reef. Reefs near shore may also be overfished, depleted of the grazing fish that eat algae off a reef and the predators that keep the ecosystem in balance.

Since their heat-tolerance superpowers will be essential as climate continues to warm, it's particularly important to protect the corals in these shallow reefs from local threats. The Super Reefs Initiative aims to do this by finding coral boot camps and connecting these reefs with conservation efforts. Marine scientist Anne Cohen leads the project and brought together her team at the Woods Hole Oceanographic Institution with Steve's team at Stanford University, conservation experts from The Nature Conservancy, and other organizations. Their project can be summed up as a three-step process: "We call it predict, prove, and protect," says Steve. They predict where super reefs are located, prove that the corals in those reefs are heat tolerant, and then work to protect these locations from local stresses.

To predict where super reefs are located, Anne's team uses ocean modeling to search the world for reefs with conditions much like the daily temperature swings of Ofu, which may allow corals to be more heat resistant. They also search the ocean model for other types of super reefs, including possible refugia where waters are cooled by natural processes like currents, tides, and internal waves (as we explored in chapter 7).

Steve's team tests the heat tolerance of corals in the locations where Anne's team identifies daily temperature swings. Because the ocean is enormous, and the locations identified span the tropics, coral heat testing (similar to the testing done in Ofu) needs to happen in many locations. "When there's a problem in the world's biggest habitat, it's a big problem. And so, solutions have to scale up," says Steve as he explains how his group has been training people to test coral heat tolerance in the Marshall Islands and Palau, two of the locations identified. Local community college students studying environmental science are working with local scientists on these islands to test the corals. Steve's group provides them with mentoring via Zoom and the equipment they'll need. Getting local community members involved helps the project scale up, and it also allowed it to continue when travel was limited during the Covid-19 pandemic.

Once a super reef is found and tested, experts from local conservation groups and The Nature Conservancy work with communities to help protect reefs from problems like pollution and overfishing and to include the heat-tolerant corals in reef restoration projects. The hope is that someday, coral larvae from these reefs will make their way to reefs where corals haven't been able to survive the heat, where they will colonize, reviving the reef ecosystem.

~

The small and patchy cauliflower corals (*Pocillopora*) in Golfo Dulce, Costa Rica, don't look like super reefs, but they have something in common with reefs like Ofu's village of elders: daily swings in environmental conditions. In Golfo Dulce, corals are exposed to daily changes as tides bring water into and out of the narrow gulf. At low tide, water heats up and becomes more saline and acidic. At high tide, a flood of cooler water enters the gulf, giving corals a break from the extreme conditions. Raising Coral collects water temperature data at all of its sites in Golfo Dulce and has found that the corals are experiencing, and surviving, daily swings in temperature of up to 14°F (8°C), an even larger swing in temperature than in the Ofu backreef. Joanie Kleypas, Raising Coral director, credits the tidal temperature swings with helping the corals become more resilient to heat, although even with extra resilience, they are still tested by extreme events.

It's unknown what combination of adaptation and acclimatization may be responsible for the observed resilience, but cauliflower corals have several strategies that could help them deal with environmental stress in addition to genetic adaptation. They can switch symbionts for more heat-tolerant types and can raise their reproduction rate, increasing the number of colonies and genets.[78] Raising Coral staff have seen the corals in Golfo Dulce spawning monthly, which is very unusual (most corals that spawn do so only once a year). The colonies can also grow in different shapes depending on sunlight and waves, which makes them particularly difficult to identify since the same species can have different shapes.

The fragments of cauliflower coral in Raising Coral's nurseries that didn't bleach in 2020 were from colonies in water prone to temperature swings. About one out of five coral fragments in Raising Coral's nurseries died after the 2020 bleaching event, and most of those were two coral genets that were not as resilient to heat. But more than 80 percent of the coral fragments survived and were outplanted in reefs. "We're going to take the ones that are the toughest coral we have," says Joanie.

A bleaching event in 2016 delayed the start of the Raising Coral project (the same heat that killed some staghorn coral at Coral Gardens Reef, as we learned in chapter 7). It would have been futile to harvest and try to grow fragments of bleached corals, so the team waited.

"They stayed bleached for like four months, which is unheard of," Joanie recalls. Surprisingly, most of the bleached corals recovered. "There was some mortality but not anything like I was expecting," she says.

Joanie used to work on the Great Barrier Reef, with its expansive framework and high biodiversity. In comparison, the Eastern Tropical Pacific where she started Raising Coral appears paltry, with limited coral species and small patches of reef. People ask her why she would work with such boring reefs. But Joanie sees tenacity in the corals, which isn't boring at all. The tough corals of Golfo Dulce give her hope. "These may be some really good corals for survival in the future," she says. "They're resilient because they have to be."

**CHAPTER 10**

# THE ANTHROPOCENE CORAL PARADOX AND THE FUTURE OF REEFS

"Is pillar coral extinct now?" Karl Wirth asks over dinner. It's December 2022, and I'm in San Pedro, Belize, for fieldwork at Coral Gardens Reef with Karl, Lisa Greer, and Al and Jane Curran. The table falls silent for a moment, except for the sound of water lapping at the edge of the seaside restaurant and the rustle of palms above. Steps away is the backreef, and beyond it, waves breaking at the reef crest. I dig my feet into the sand below the table and think about all the life out there that we're losing.

My plane had landed in San Pedro about an hour before dinner. I was excited to see everyone and meet the corals that have survived hard times in the refugia the team discovered.[1] Later in the week, I would get to see how restoration projects were spreading resilient corals into degraded reefs, which we visited in chapter 8. The ability of corals to persist and bounce back is the theme of this trip, so I temporarily forgot about all the corals that are struggling and on the brink of extinction. When you start seeing the glass as half full, it hurts that much more when you realize it's half empty. But both are true.

On December 9, 2022, the day before that dinner in San Pedro, news broke that pillar coral (*Dendrogyra cylindrus*) had moved into the "critically endangered" category on the IUCN Red List of Threatened Species, one step above extinction.[2] This is probably what Karl heard, and this news may have been combined with reports that pillar coral had become functionally extinct on Florida reefs,[3] meaning that the small scattering of living colonies (a total of 40 are known in Florida waters[4]) are no longer enough for the population to reproduce in that area.[5] Throughout the Caribbean region, the only place they live, the number of pillar coral colonies has dropped by more than 80 percent since 1990.[6] The species tends to be particularly susceptible to disease. Heat stress and low water quality make it even more vulnerable.[7]

Three years before we were at the seaside restaurant, Al and Jane Curran had seen a large and very healthy pillar coral in the backreef near San Pedro. From our dinner table on the shore, Al recounts it, pointing out to sea northeast of the restaurant, in the general direction of the coral colony. Lisa and Karl remember it too, in shallow water offshore of Amigos Dive Shop. Sharing stories about the pillar coral, they sound like people who have just discovered they have a friend in common.

Often pillar coral is just a few vertical towers that connect at the base. It may look like an oversized glove, its fingers extending upward as if it's raising its hand to ask a question in class. But the specific pillar coral they were describing was a sprawling colony with an underwater skyline of pillars. In Al's photo of the coral, I count over a hundred pillars with rounded tops growing toward the sunlight, rising high above the surrounding seagrass, each tower more or less cylindrical. The coral's light brown surface looks as fuzzy as a Muppet. That fuzz is actually the tentacles of the pillar coral polyps waving in the water. Unlike most corals, pillar corals are usually found with their tentacles extended and swaying as water flows past.

Pillar coral's fuzzy surface makes it very recognizable, so it makes sense that they would all take note of the colony. And a pillar coral with so many pillars is unusual, and thus noteworthy.

It would be odd to know the specific whereabouts of rare species that can move around—a particular shark or salamander, for example—unless they had been tagged and were tracked by satellites. But because coral stays put on the seafloor, the researchers knew exactly where we could find this pillar coral colony the next day. It has been a fixture in a neighborhood of scattered patch reefs, seagrass beds, and rippled sands that is familiar to the four of them who have been studying reefs in this area for years.

After dinner and a night's sleep, and after collecting data along transects at Coral Gardens while being tossed in the waves much more than my stomach could handle, Lisa and Al describe the giant pillar coral with over a hundred pillars to our boat captain Narciso Valdez. Amazingly, he knows what they are talking about and steers the boat right to it. Everyone, apparently, knows this pillar coral.

Once Narciso stops the boat, Al and Lisa jump into the water to look for the coral. They snorkel around for a couple of minutes, and then Lisa dives down, having found the right spot. She resurfaces, pops her head out of the water, and pulls the snorkel out of her mouth. "It's like an abandoned city down there," she shouts to the four of us who are still on the boat.

The pillar coral is entirely dead. It's now a gray skeleton splotched with encrusting algae and other organisms. The skeleton looks battered, the tops of the tall pillars broken off. Lisa describes it as a dystopian cityscape. Its skyscrapers look vacant and blank, although they have probably become home to bioeroders and encrusting life (like those we met in chapter 4) that are chipping away at the skeleton and covering it up.

"It looks like someone went in with a meat cleaver," says Al after he climbs back aboard. He isn't being literal, just referring to the broken pillars, but a gruesome image comes to mind—a coral massacre. Of course, that's not at all what happened. The most likely culprit that killed the enormous pillar coral is disease—stony coral tissue loss disease, which arrived in Hol Chan Marine Reserve in 2020, six years after it was first found in Florida. The second most likely culprit is heat.

Despite many corals demonstrating their superpower to persist, the plight of pillar coral is a reminder that not all corals have similar abilities. Some are sitting ducks. And even coral with resilience can reach its limit, as we would see a year later when the staghorn coral in Coral Gardens died in the heat.

∼

The giant pillar coral off San Pedro, Belize, alive in 2019 (top) and dead in 2022 (bottom). Both photos by H. Allen Curran, Smith College.

Nearly 200 years ago, Charles Darwin described a coral reef paradox: the ocean's most biodiverse ecosystem seemed to exist in its most nutrient-poor areas.[8] That's a fine paradox for the nineteenth century, when the ocean was thought to be too large and powerful to collapse, but not today. As we now consider whether reefs will be able to survive the Anthropocene, another paradox comes to mind: that of the corals themselves, which can be both amazingly resilient and terribly vulnerable.

Certain coral species or genets might be more resilient than others, which is great because it means that some may survive climate warming and local hazards, but even tough corals that have survived for centuries can perish in days, weeks, or months. As I have researched this book, I have encountered many examples of how some corals have shown the ability to persist over years and decades of difficult times and how the broader reef ecosystem has shown the ability to persist over thousands to hundreds of thousands of years. These examples make me hopeful that corals may be able to use the same abilities and adaptations to persist into the future. And yet I've also encountered endless examples of corals that have been decimated, like the giant pillar coral, after centuries of resilience.

"They're delicate flowers because everything we do kills them," says Steve Palumbi when I express my astonishment at how corals can be both powerful and vulnerable. "But when we just give them the slightest chance, you know, they kind of begin to grow back—not all of them, but they kind of begin to grow back."

Fossil reefs that lived before the Anthropocene, like those that hug the coasts of many tropical islands, illustrate the ability of reefs to endure over the long term, with similar patterns of reef zones and the same coral species filling both Pleistocene fossil reefs and healthy modern reefs before the stresses of the Anthropocene.[9] Before their recent changes, the ecology of living reefs mirrored that of fossil reefs, so researchers could use the principle of uniformitarianism to interpret fossil reefs, as we saw in chapter 3. The same coral species returned again and again to build reefs that are fossilized in stair-step terraces on Barbados and Papua New Guinea as the land uplifted.[10] In the long term, this has left a long record of reef recovery.

Corals were killed and reefs were decimated by environmental changes before the Anthropocene. But once conditions improved, the fossils show that reefs rebuilt, reliably bouncing back from catastrophe over the long term. Sometimes it took decades to a century—as it did for the Papua New Guinea reefs we visited in chapter 5, which were killed by volcanic ash and other acute disturbances. Sometimes it took a thousand years or more—as it did for the fossil reef at Devil's Point, decimated by a wiggle in sea level, and reefs off the Pacific coast of Panama that stopped building as ENSO grew variable, as described in chapter 6. Nearly all coral species survived large changes in sea level during ice age cycles, presumably sheltering in refugia when needed. Where the Great Barrier Reef is today, there's evidence that reefs formed, died, and then formed again five times over 30,000 years, the species lying low between reef-building periods like actors waiting just offstage for their cue.[11]

Evidence of refugia like Coral Gardens, which persisted through the Caribbean reef declines of the twentieth century, is a hopeful sign that corals can still persist in some places.[12] Because environmental conditions aren't the same everywhere, there may be pockets where reefs can continue to survive.

And yet we've also seen corals die quickly when environmental conditions change, when a new disease debuts, when water quality and fish numbers drop, and when ocean water overheats as it did in 2023, toppling the Coral Gardens refugium. A third of coral species worldwide are at risk of extinction,[13] one part of the global mass extinction that's been unfolding because of human activity.[14] Where corals die off, other reef life that relies on coral for habitat suffers.

Currently, climate warming is causing more marine heat waves—periods when ocean water is unusually hot—which leads to more coral bleaching. Widespread bleaching events were unheard of before 1979 but are becoming increasingly common.[15] The 2023 heat that killed the staghorn coral at Coral Gardens would become the fourth global bleaching event, which was declared in April 2024 as the heat spread worldwide. Previous global bleaching events include the third global bleaching event, which occurred over an extended timeframe from 2014 to 2017;[16] the second global bleaching event, in 2010; and the first global bleaching event, in 1998.[17] Before 1998, coral bleaching occurred (including disastrous mass bleaching in the 1980s), but not at a global scale.

All four global bleaching events occurred during El Niño, although the third and fourth events started before El Niño.[18] While El Niño is a natural cycle, its effects are amplified by climate change, causing stronger marine heat waves. Because the El Niño between 2014 and 2017 lasted so long, it was particularly destructive to reefs, causing the third global bleaching event to have the most widespread coral bleaching ever recorded[19] (but it's likely that this record will be broken as climate continues to warm).

Watching how corals and other reef life were affected by heat during the third global bleaching event gave researchers a glimpse of how future events, supercharged even more by climate change, may affect coral reefs. They found that coral bleaching also affects other reef life like fish and invertebrates,[20] and they learned how the coral microbiome reacts in hot water, and how corals can swap their zooxanthellae for more heat-tolerant varieties.[21] In some reefs, scientists found that bioerosion outpaced limestone production during the event, as limestone production slowed when corals bleached and died and the bioeroders chewed apart the structure of reefs much faster than expected.[22] Research on the bleaching event also helped scientists see a connection between coral heat stress and disease.[23]

Although we don't know exactly the role that heat played in the disease's impact, it was during the third global bleaching event that stony coral tissue loss disease (SCTLD) first spread through Florida reefs and eventually through the Caribbean. We have a detailed account of the disease onset because there happened to be a team of scientists closely monitoring corals near Miami in 2014 at what would become ground zero for SCTLD. Over two years, they watched corals to document any impacts from nearby dredging at the Port of Miami. For comparison, they also monitored corals at sites that were too far away to be affected by dredging, called control sites. That way, if a problem affected all the corals, they would know that it wasn't caused by the dredging. For example, in August 2014 nearly all the corals, at both the dredging sites and the control sites, bleached, which was caused by overheated water in the region, not dredging. The corals were still bright white at the end of September when the team found three colonies at a control site with a strange unknown condition. Comparing photos of these three corals from week to week, they noticed a zone of dead coral polyps on each.

Once the water temperature cooled enough in the fall that corals could regain their zooxanthellae and color, the effects of the new mystery disease became clear—the parts of each coral that had been killed by the new disease stayed white because only the skeleton remained.

Reef geologist and ecologist Bill Precht, whose consulting firm ran the coral monitoring project, jumped into the water at the control site late in the fall and was shocked to find that about half the corals were either recently dead or diseased.

"I got back to the boat and I was like, oh my god, I've never seen anything like this. And I've seen white band disease," Bill recounts. Seeing a new disease take hold gave him flashbacks to his experience watching the reef at Channel Caye, Belize, die from white band disease in the 1980s. "I was sick to my stomach when I got back to the boat. I mean, it was horrific."

Once the disease spread beyond their study area, Bill and two of his daughters (one had just finished college and the other was in high school) would dive about every other weekend in the shallow waters off Florida, following the disease as it spread, helpless to do anything to stop it. In the year after it was first found, SCTLD spread about 80 miles (130 km) along Florida's east coast.[24] It's since spread throughout the Caribbean and has proved to be difficult to stop, adding another challenge to the suite of ongoing problems facing reefs in the Anthropocene.

Corals are no strangers to coping with change on multiple timescales. Each day they are faced with tidal cycles that can send the shallowest corals into overheated water, and these temperature changes can help build their resilience to heat, as we saw in the previous chapter. Each year they are subjected to seasonal changes in water temperature, with the highest chance of bleaching in the late summer and fall. Every few years the balance between El Niño and La Niña shifts, which changes the patterns of weather and ocean temperature. Over tens of thousands to hundreds of thousands of years they have been subjected to changes in sea level with ice age cycles.

These changes, which have affected both contemporary corals and now-fossilized corals, are cyclical. The changes we are causing in the Anthropocene—the warming climate, acidifying oceans, strengthening hurricanes, overfishing, pollution, and declines in water quality—are not. There is no precedent for this in the history of coral reefs before the Anthropocene; the fossil record is of no use to us here.

It's predicted that most reefs will be lost this century because of these various Anthropocene stressors.[25] Such stressors don't have the same pattern as the cyclical changes that corals typically endure. These one-way changes are neither cyclical nor sustainable. They can affect the cyclical changes, for example by making summer heat and El Niño more extreme and leading to the unprecedented heat in 2023 in the Caribbean and Eastern Tropical Pacific.[26] For reefs to survive, we need to put a stop to the one-way changes.

Several of these changes are a direct result of the carbon dioxide and other greenhouse gases that we are spewing into the atmosphere. These gases cause the climate to warm by trapping heat. The most prevalent greenhouse gas, carbon dioxide, dissolves into seawater, which is making the ocean slightly acidic—a challenging condition for corals and other limestone-producing marine life. And as greenhouse gases warm our climate, marine heat waves that cause coral bleaching are becoming more common, and hurricanes are growing stronger, with faster winds and more extreme rainfall, increasing the chances of disaster for coral reefs in the storms' paths.

There's a curious connection between fossil fuels (the source of most greenhouse gas emissions[27]) and coral reefs. Of the conventional oil and gas that remains underground, over 60 percent of the world's oil and 40 percent of its gas are in limestone and similar rocks that typically form in shallow tropical seas, according to estimates by global drilling giant Schlumberger.[28] These rocks are millions of years older than the limestone platforms we've encountered in this book, but they formed in much the same way, built over time out of the skeletons of dead marine life. The oil and gas within the rocks formed from the dead bodies of animals and algae that lived in reefs and elsewhere on shallow limestone platforms. Often these were species that are now extinct, but they had roles similar to those of tropical marine life today. The bodies of dead reef life that were buried (instead of decomposing or being eaten on the seafloor) were under pressure and heated for millions of years, which transformed the dead into oil and gas, fluids that slip through the spaces between sand grains and fossils like underground ghosts. Burning these fossil fuels is, in part, burning the remains of reefs, which releases carbon dioxide and causes problems for reefs today. It's striking that one of the main ways we are currently harming coral reefs is by burning their ancestors.

The Paris Agreement, which was signed by nearly 200 nations plus the European Union, agrees to limit greenhouse gas emissions enough to keep the total amount of human-caused climate change to under 2°C (3.6°F) while also making efforts to reduce emissions enough to keep the amount of warming closer to 1.5°C (2.7°F).[29] If we meet these goals, it's likely that many reefs will survive. If we can limit temperature rise to 1.5°C, between 10 percent and 30 percent of coral reefs will survive, according to the IPCC.[30] Presumably, more corals will survive in reef refugia while other locations with less livable conditions will lose all coral cover. The hope is that once climate is stable (albeit warmer, perhaps for centuries), corals that have survived the heat will be able to spread into reefs where corals died.

However, there may be a big difference between 1.5°C and 2°C warming in terms of the number of corals that can survive. If the climate warms to 2°C by the end of this century, only 1 percent of coral reefs may survive, according to the IPCC.[31] Each fraction of a degree added to global average temperature will likely make a big difference for reefs.

Another study looked at how future limestone production in reefs may be affected by climate change and ocean acidification at 183 reefs worldwide. It found that with 1.5°C to 2°C warming this century (the goal of the Paris Agreement), almost two-thirds of reefs will still, at the end of the century, be able to make more limestone than is eaten away by bioerosion.[32] But with higher levels of warming and acidification, most reefs will not be able to build limestone faster than it is bioeroded. These reefs will be unable to keep up with sea level rise and will be at risk of drowning over thousands of years as they become deeper, eventually beyond the reach of sunlight.[33]

The equation changes somewhat when we consider corals' ability to adapt and acclimatize to heat. Although there's a lot we don't know about the degree to which corals can adjust to cope with warming, modeling shows that slowing the rate of warming in line with the Paris Agreement goals will give many corals the time they need to adjust (assuming they have the ability).[34] With 1.5°C to 2°C warming, and corals with some ability to adapt and acclimatize, there would likely be somewhat more coral bleaching and death until about midcentury, and then the situation would likely stabilize as corals become more resilient and warming slows.[35]

There isn't time to spare if we are to meet the Paris Agreement goals. But we are not on track to meet either the 1.5°C or 2°C goal, according to a 2023 United Nations Emissions Gap Report.[36] To limit warming this century to 2°C, we need to cut greenhouse gas emissions by 28 percent by 2030, and to limit warming to 1.5°C we need to cut emissions by 42 percent by 2030, according to the report.[37] This will require large-scale transformations of our energy sources away from fossil fuels, but even small emission reductions add up, and each small reduction in warming makes a difference for life on land as well as corals and other marine life. According to Inger Andersen, executive director of the UN Environment Programme, "Every fraction of a degree matters: to vulnerable communities, to species and ecosystems, and to every one of us."[38] Of course, saving coral reefs is just one of a multitude of reasons that we urgently need to solve climate change.

While there is no easy fix to slow climate change to levels that will keep coral reefs from collapsing, there are a lot of actions we can take to get to that point. Diving into the details of all that we can do to solve climate change is beyond the scope of this book, but in general, two types of actions are needed: we need to decrease the amount of greenhouse gases like carbon dioxide that we add to the atmosphere, and we need to increase the amount of carbon dioxide that's taken out of the atmosphere.

Since the majority of greenhouse gas emissions come from burning fossil fuels, decreasing emissions means a switch to energy sources that don't have emissions, particularly renewable energy like wind and solar energy. Technology is in development to harness renewable energy from the ocean too, such as underwater turbines anchored to the seafloor that may soon turn tidal energy into electricity, and floating buoys that capture wave energy as they bob up and down.

To decrease the amount of carbon dioxide in the atmosphere we need to expand and protect carbon sinks that pull carbon out of the atmosphere while we reduce emissions. The oceans could help with this while also providing other benefits for coral reefs. For example, mangroves that grow near coral reefs and are home to many juvenile reef animals are particularly efficient at trapping carbon dioxide,[39] and so are seagrass beds, another reef-adjacent habitat where species common in reefs often spend part of their lives.

On a local scale, taking action to help reduce chronic problems—like water pollution and overfishing—helps reefs build resilience so they have more capacity to bounce back from disturbances like warmer-than-usual water and hurricanes supercharged by climate change. Efforts to keep fish populations healthy around the island of Bonaire in the southern Caribbean may have helped the area's reefs recover from a phase with lower coral cover and more macroalgae, which was caused by bleaching and a hurricane.[40] With enough parrotfish to graze the macroalgae, corals could return.

Marine biologist Nancy Knowlton, recognizing that all the bad news about marine degradation was not inspiring helpful actions, started to chronicle the good news about the ocean in an online campaign (#OceanOptimism), presentations, and a 2021 article titled "Ocean Optimism: Moving beyond the Obituaries in Marine Conservation."[41] She highlights examples of actions that are helping oceans, including reefs—for example, how protecting marine areas helps fish populations grow. One study found almost seven times more fish (measured in biomass) in protected areas that don't allow fishing than in unprotected areas.[42] And when protected areas of the Great Barrier Reef were expanded in 2004, fish became more abundant and there were fewer outbreaks of coral-eating crown-of-thorns starfish.[43] There's also been some growth in the populations of protected species like sea turtles. For example, after the United States protected green turtles in 1978, the number of turtles increased steadily in Hawaii, and green turtle populations increased significantly in the US Virgin Islands and Florida after about 2000.[44] Perhaps we can do the same for other reef life.

For water to stay clear of pollution like sewage, sediment, plastics, extra nutrients, and pesticides, activities on land along the coast and in the watersheds that feed into shallow tropical waters need to be managed. Preventing runoff from agriculture and coastal development can help reefs avoid becoming dominated by macroalgae, having corals smothered by sediment, or even suffering from outbreaks of disease. While there is some good progress on these local-scale efforts—enough that we know they work—not all regions have protections for reefs and systems for wastewater and runoff.

Restoring reef corals to actively bring back what's been lost is also effective. Reef restoration without solving the climate crisis will not work long term, but restored reefs can help ensure that there are corals to repopulate reefs when we do slow warming. After Lisa Greer, Karl Wirth, and I visited the restored reef at Laughing Bird Caye, Lisa put it this way: "We're past the days of letting nature just be nature. I mean, if we want coral reefs, then we may need more reefs like Laughing Bird that are engineered."

Will all our efforts ensure a future for coral reefs? There's a lot we don't know, but we do have a good idea that these strategies of limiting warming, improving local conditions, and restoring reefs will work. Even with all these efforts, not all corals will make it into the future, but enough can be saved to let reefs persist.

"I think some of them will make it. Some of them will not, but I do believe that a lot of corals are [heat] resistant enough to make it," Steve Palumbi tells me when I pepper him with questions about the future. He's witnessed corals' abilities to adapt and acclimatize, as we saw in the previous chapter, but he knows the near future isn't rosy. Still, he sees that corals have the ability to survive as long as we can limit climate change and improve local conditions. "Over the next, you know, 60 or 70 years, as things get bad, really bad, there'll be enough coral generations to spawn off ones that are even more resilient than we have now."

Coral reefs have not been totally lost, according to John Pandolfi, whom we met earlier. "I think the trajectory is depressing, and erodes hope, but currently we're not at that sort of hopeless stage yet," says John when I tell him that I get questions about whether it's too late to save reefs. The news headlines about disastrous mass bleaching and other reef calamities are so dire that it's easy to conclude that we missed the window to help coral reefs survive. "There's a lot out there that's still worth saving and worth working for," John says. "And even though there's a lot that's been lost, and my children will never see what I saw, and I'll never see what my great-grandparents would have seen, there's still a lot out there to appreciate. So, I'm hopeful that reefs will keep going. And I think that sort of sustains me."

We are causing problems for corals today that have never happened in the past, testing their ability to bounce back. Some corals appear to be up to the challenge. Some do not. And even corals that have some resilience can hit a tipping point where there's just too much stress. Most corals likely won't survive this century, but that doesn't mean that coral reefs will be lost forever. There's a big difference between almost doomed and totally doomed. The former has the ability to recover. There's a lot we can do to make sure coral reefs are not totally doomed, and to help them bounce back. The situation is dire, but there is reason for hope.

∼

Someday the only evidence of the current Anthropocene tumult will be in the fossil record. To get some perspective on what could happen, let's imagine the current world as if it were the distant past, as if we are in the future and have the benefit of hindsight. Imagine we are looking at a rock outcrop on the edge of a tropical island that preserved the events that occurred before, during, and after the twenty-first century. Thinking about the shallow end of deep time can help us see what the road ahead could be for reefs and how the story of reefs in the Anthropocene will read in the fossils and subfossils. Maybe, like the Cockburn Town fossil reef on San Salvador, Bahamas, the ocean is lapping up on the seaward edge of the fossil reef. Or maybe sea level is higher and we are coring below the seafloor and below the corals that we hope are living there, gathering fossil or subfossil evidence of what happened from the layers of limestone.

What reef stories might be preserved in the subfossils and fossils on the layer cake limestone platforms over the next decades and centuries? We are currently in the middle of the mess, and it's going to get even messier over the next few decades, but taking this long view can remind us that while we are reading headlines about bleaching events, coral diseases, and other reef problems, we also need to focus on the long-term goals of stabilizing the climate and helping reefs recover eventually. Reef recovery might not happen in our lifetimes. Evidence from the past shows that it could take at least a century, perhaps even thousands of years. It can be hard to act when we are not personally going to see the end result: healthy reefs in a distant future. But the actions we take today will determine whether reefs recover.

There will be a time after we stop releasing so many greenhouse gases. If we can slow the warming soon and protect as many reefs and species as possible, there's a good chance that corals will eventually return, although perhaps not all coral species.

Our actions may leave gaps in the reef fossil record. For example, if we lose all but a fraction of coral reefs in the coming decades, in many places corals won't actively build reef limestone for centuries or even millennia, or they will build limestone slowly and it will then be eaten away by bioeroders. Because macroalgae don't produce limestone that can fossilize, it might be impossible to see where algae outcompeted corals, where reefs underwent a phase shift, or which corals died of disease, although we might be able to recognize this by looking at taphonomic alteration of the coral skeletons in the fossil record, as we saw in chapter 4. These declines in reef corals would leave an unconformity in the rock, a plane through the limestone that belies the drama of what happened on Earth during this turbulent time. It might look like the unconformity in the fossil reef at Devil's Point, and somewhat like the unconformity that James Hutton saw in a rocky coast of Scotland that helped him understand the immensity of geologic time and see that the present can be the key to the past.

If, in the future, you were to look closely at the unconformity that reflects the current time, you would not see much, or perhaps any, evidence of the actions people are taking. You would see no evidence of divers monitoring reefs and applying antibiotics to diseased coral colonies. You wouldn't see evidence of governments above the sea surface making legislation to protect parrotfish and the shallow marine world of reefs. You may not see renewable energy infrastructure built to replace fossil fuels. Overall, you wouldn't see much evidence of all the actions people are taking now that will determine whether reef life will be able to squeak through these hard times.

During that unconformity, when many areas are without reefs, we humans would be suffering too. We depend on reefs. Coral reefs support economies through tourism and fisheries.[45] The ability of coral reefs to protect coasts is increasingly valuable as landfalling storms are supercharged by climate change. Reefs currently protect more than 43,000 miles (70,000 km) of coastline, buffering more than 90 percent of waves and reducing the amount of coastal storm flooding by more than half (which costs US$4 billion a year).[46] Reefs may also hold important medical breakthroughs and even cures to diseases. Communities that live near reefs often rely on them for food and have cultural connections to the ecosystem.[47] Helping as many reefs survive as possible doesn't only help ensure a future for reefs—it also helps us.

The layer that forms above the unconformity is where you would find the results of our efforts to help. This layer may include fossilized reefs if we successfully limit climate change and help species survive in refugia until conditions improve. There is uncertainty about that layer of limestone. You might find that it formed after hundreds or even thousands of years. You might find that it has fewer coral species than the older fossilized reef below. With so many corals facing extinction, it's likely that they wouldn't all make it. Maybe it would hold fossils of heat-tolerant corals from the Red Sea or Ofu in American Samoa if local coral species could not survive.

In some places, there might not be an unconformity in the limestone. Where reef refugia could persist, you might find a story of continuous reef growth in the fossils, even if the ecosystem had to cope with bleaching and other disruptions. Perhaps this is what the coast of Ofu would preserve or the northern Red Sea. If these refugia are protected and fish populations rebound, you may also find that the number of parrotfish teeth in this layer of fossil reef increases over time.

This patchiness of reef health means that there are still bright spots amid the present-day chaos. In Belize, healthy staghorn coral that survived twentieth-century decline shifted my baseline for what a Caribbean reef could be like (although that coral, which survived so many other threats, wouldn't survive the 2023 heat). Marine scientist and science writer Juli Berwald, in a 2023 *Nautilus* article, described reefs of Tela Bay, Honduras, that are surprisingly healthy despite murky waters from runoff.[48] And in a 2022 *National Geographic* article, explorer-in-residence Enric Sala described his shock when he first jumped into the shallow water of the Line Islands in the middle of the Pacific and found a "thriving coral jungle full of large fish" and then, over a decade later, more shock as he saw on a return visit how the reef had bounced back after the third global bleaching event.[49]

The fact that we are now shocked to see a healthy reef reveals the bleakness of the situation, but it also means that all hope is not lost. Worldwide, the search for still-thriving reef refugia is ongoing. Are these refugia enough to help species survive? Perhaps, but it's uncertain. What is certain is that the more reefs we can protect in the next few decades, and the more we can slow climate change, the more likely it is that corals and other reef species will survive.

In some locations, the fossil record might have an unconformity that represents only a few decades, with a reef preserved above it that was restored, its corals grown and tended by people after the previous reef died. Maybe some of the nails and cement used to attach new corals to a restored reef would be preserved alongside the fossils. Maybe there would be evidence of coral nurseries in the fossil record too, with table legs jutting out from between fossil corals, the ropes strung across them probably not preserved in the rock. Take a close look at the fossil coral skeletons within the limestone of a restored reef, and you might find that the corals changed over time as people selected, grew, and outplanted the most well-adapted corals, a version of artificial selection in the wild. If you are looking at a Caribbean fossil reef that was restored, you might see fossils of staghorn and prolifera corals that grew on the unconformity and then fossils of massive coral species above them in the fossil reef, arriving only once the disease attacking them (SCTLD) was under control.

When reef geologist and ecologist Lauren Toth's research brought her to a Florida reef where coral restoration experiments were in progress, she saw tests of how different massive coral species grow when outplanted on top of dead corals in a reef. "There were some of these dead coral heads that had five different species of massive coral in different clusters," Lauren recounts. This would make for a strange fossil record with the skeletons of different coral species stitched together, but it's a good reminder that reef corals live among the skeletons of their dead ancestors. "If you cored through that, you would be very confused," says Lauren, considering the record that the corals might leave behind.

You might also find that sea level rise caused the future fossilized reef to form on what is land today, especially if it grew after a thousand or more years of sea level rise. On San Salvador, Bahamas, this may create a layer of fossil reef on top of the Pleistocene Cockburn Town fossil reef. There would be an ancient soil (a paleosol) at the unconformity, indicating that this was land above sea level. The remains of Cockburn Town's colorful buildings and giant iguana statue might be preserved as well, including walls made of blocks of fossil reef taken from the quarry, a striking clue to how sea level rise affected people. In the tropics, coastal communities are particularly vulnerable to sea level rise. More than half of the world's low-lying coasts are in the tropics.[50] Researchers estimate that by the year 2100 over 400 million people worldwide will be living with coastal flood risk, including about 295 million people in the tropics, which will make some areas uninhabitable and force millions to leave their homes.[51] People in flooded communities may be able to persevere, moving to higher ground as sea level rises, much as corals moved lower to stay underwater when sea level fell in the past. It will be challenging and costly, but some towns and cities may have no choice but to move as sea level continues to rise.

The situation is grim and yet there's also hope for reefs. Once climate change is under control, species in refugia can spread out and form thriving reefs, as long as local marine environments are healthy. And those reefs may leave fossils behind to be found by people thousands or even millions of years from now. Perhaps a scientist exploring the fossils on a tropical coast will look down someday and see sinuous lines in the limestone below her feet, the skeleton of a brain coral that survived the Anthropocene.

When I began writing this book, I was focused on the seemingly healthy reefs in the Pleistocene and Holocene fossil record compared with the often-abysmal health of reefs today. On the surface it seemed straightforward: reefs were healthy without us and now are less healthy with us. We humans are just terrible in the Anthropocene, causing extinctions, polluting the oceans and land, creating messes everywhere. Chances are there will be a lot of plastic in the limestone from this time.

But as I explored the research and talked to experts, I realized that, like corals, we humans are also a paradox. We cause numerous problems for reefs, and yet we can also be innovative and solve those problems. Someday, the story of how we helped coral reefs might be captured in the fossil record, or the story of how we destroyed this ecosystem will be preserved. I hope it will be the former, not the latter. We are destructive, yes, but we also have a remarkable ability to fix things.

~

Several days after Lisa Greer and Al Curran checked on the enormous pillar coral near San Pedro and found only a dystopic scene of dead limestone pillars, I would see another pillar coral that was very much alive. This was in southern Belize at Laughing Bird Caye while I was snorkeling with the Fragments of Hope team. As I kicked my flippers across the restored reef's tangle of staghorn corals, tiny fish dove into the space between branches. I swam by larger fish parked in the shade of elkhorn coral canopies. And I carefully crossed over broad expanses of shallow prolifera to ensure I didn't touch the corals. It was hard not to have hope for the future of reefs.

As I swam beyond the thickets of branching corals, a coral with a distinctly different shape caught my eye. It was on the outskirts of the restored reef, a brown bulbous mound of coral surrounded by the drab

A wild pillar coral near Laughing Bird Caye, Belize, with staghorn coral from the restored reef.

death assemblage, cobbles of reef corals that died decades ago, and sparse living colonies of staghorn coral, a yellow color and about the size and shape of tumbleweeds. Reaching up from the depths, it was a sizable pillar coral with a number of stocky pillars. Every inch of it was alive, with the long tentacles of coral polyps swaying back and forth with the waves. It wasn't nearly as large as the pillar coral colony off San Pedro that fell into ruins, but it was large enough to fill me with hope for the species. The light filtering through the shallow water struck the tentacles of the coral polyps and made the colony surface look even fuzzier than usual, like a brown fleece blanket had come to life.

When faced with an endangered species that I may never see again, my instinct was to get out my camera and document every inch of it. I was not the only one. Lisa, Karl, and I surrounded the lumpy colony like paparazzi taking photos and videos, but unlike actual paparazzi, we kept our distance, not wanting to harm this delicate flower.

According to Lisa Carne at Fragments of Hope, this pillar coral is natural, not intentionally added to the restored reef. In her words, it's a wild coral. And so far, it has somehow avoided SCTLD. There are a few other pillar corals near Laughing Bird Caye, too. Our group would spot three in total during our snorkel around the island.

If there's still hope for pillar coral to survive extinction, it's largely because of conservation efforts that may someday lead to a sustainable pillar coral population in the Caribbean. Restoration projects are underway to grow it from fragments, which can be a struggle because of its susceptibility to disease but sometimes yields good results. A few pillar coral colonies were still alive in one of Fragments of Hope's Belize nurseries after SCTLD swept through. In Florida numerous organizations are working to save pillar coral by growing colonies in tanks on land. The Frost Museum of Science in Miami is temporary home to more than a hundred pillar coral fragments collected from Florida reefs.[52] Living on land and indoors, not far from the park benches made of fossil reefs, they are biding their time. One day, these pillar corals will return to the sea.

Whatever the future holds for reefs, our actions will be a huge part of their survival or their peril. What we know about reefs in the past can guide these actions to help reefs in the present, in hopes that they eventually have a vibrant future. Even if that goal is beyond our lifespans, there's a lot we can do. Our efforts to help reefs may seem limited, but they collectively have an impact. Just as it takes large numbers of polyps, zooxanthellae, and other microbes to build a coral colony, and large numbers of corals, fish, sharks, urchins, and other organisms to build a reef, it will take large numbers of people to help reefs survive in an increasingly inhospitable world.

## ACKNOWLEDGMENTS

Thank you to Alison Kalett, Hallie Schaeffer, Kathleen Cioffi, Laurel Anderton, and the whole team at Princeton University Press for working with me to make *Reefs of Time* as polished and thoughtful as possible. And thanks to my agent, Jessica Papin, who found such an excellent home for this book.

I very much appreciate the insightful feedback of two reviewers, Anthony Martin and Willem Renema. I'm also very thankful to Adam Holloway, Mary O'Reilly, and Julie Malmberg for notes and edits on the manuscript when it was still very much a work in progress.

I'm incredibly grateful to the scientists who were so patient and generous with their time as I peppered them with questions about research above and below sea level, especially Scott Bachman, Paul Blanchon, Nadia Bood, Lisa Carne, Katie Cramer, H. Allen Curran, Kirah Forman-Castillo, Ben Greenstein, Lisa Greer, Joanie Kleypas, Loren McClenachan, Heather Moffat McCoy, Jennifer McWhorter, Steve Palumbi, John Pandolfi, Bill Precht, Lauren Toth, Sally E. Walker, Mark Wilson, and Karl Wirth. I'm also grateful to the hundreds of other scientists whose work I describe from their published research, and to Karl Wirth, Sally Walker, and Al Curran, who kindly allowed their photographs to appear in this book.

As a welcome distraction from writing this book, I wrote a feature for *bioGraphic* magazine about Belize reef refugia and restoration, parts of which are now in this book as well. The editorial expertise of Krista Langlois, Steven Bedard, and others on the *bioGraphic* team helped the article shine, which no doubt benefited the book as well. I'm forever thankful to Lisa Greer, Karl Wirth, and Al and Jane Curran for welcoming me into their Belize field research, to boat captain Narciso Valdez for getting us to the reef, and to Kirah Forman-Castillo, Lisa Carne, Dale Godfrey, Amir Neil, and Victor Faux for sharing their reef restoration projects while we snorkeled through reefs. Those experiences helped crystallize my thoughts about refugia, reef history, and shifting baselines, which enriched this book in ways I had not expected.

Several of the stories in this book draw upon my research experiences in the Bahamas, made possible with grants from The Explorers Club, Sigma Xi, the Geological Society of America, the Conchologists of America, the Latin American and Caribbean Studies Institute at the University of Georgia, the Levy Memorial Fund, and the Wheeler-Watts Fund at the University of Georgia, and a grant from the National Science Foundation to Sally Walker. Fieldwork logistics in the Bahamas were made much more manageable thanks to Sally Walker, Steve Holland, and others who joined me in the field as well as the staff at the Gerace Research Centre on San Salvador and Henry Nixon at Inagua National Park on Great Inagua.

I'm also thankful that Ben Greenstein and John Pandolfi pulled me into their NOAA-funded project on reefs in the Florida Keys years ago and introduced me to coral life and death assemblages and the stories they tell.

Closer to home, libraries also made this book possible, including the NCAR library, the Boulder Public Library, and the Denver Public Library. All three provided access to materials that I needed for research and are inspiring places in which to work.

Many thanks to Adam Holloway for his unwavering support during all parts of this project and to encouraging friends like Sara Gardiner, Moira Kennedy, Julie Malmberg, Susan Heffron, Jen Christiansen, Joel Tolman, Jonathan Foret, the crew at the UCAR Center for Science Education, and fellow writers including Christie Aschwanden, Mary O'Reilly, Jennifer Frazer, Debbie Gale Mitchell, and Mark Easter, who helped me see the path ahead for this project.

# NOTES

### Introduction

1. Ian C. Enochs et al., "Coral Persistence despite Marginal Conditions in the Port of Miami," *Scientific Reports* 13, no. 1 (April 25, 2023): 6759.

2. "Coral Morphologic Presents Coral City Camera," accessed January 27, 2024, http://www.coralcitycamera.com.

3. "Corals: NOAA's National Ocean Service Education," NOAA, accessed June 5, 2023, https://oceanservice.noaa.gov/education/tutorial_corals/.

4. "Earth from Space: Great Barrier Reef," European Space Agency, accessed July 6, 2023, https://www.esa.int/Applications/Observing_the_Earth/Earth_from_Space_Great_Barrier_Reef.

5. Nancy Knowlton et al., "Rebuilding Coral Reefs: A Decadal Grand Challenge," International Coral Reef Society, July 2021, https://coralreefs.org/publications/rebuilding_coral_reefs/.

6. Timothy McIntyre and Philip Fralick, "Sedimentology and Geochemistry of the 2930 Ma Red Lake–Wallace Lake Carbonate Platform, Western Superior Province, Canada," *Depositional Record* 3, no. 2 (2017): 258–87.

7. "Geology of Guadalupe Mountains National Park," US Geological Survey, accessed June 5, 2023, https://www.usgs.gov/geology-and-ecology-of-national-parks/geology-guadalupe-mountains-national-park.

8. Paul J. Crutzen, "Geology of Mankind," *Nature* 415, no. 6867 (January 2002): 23–23.

9. Alexandra Witze, "Geologists Reject the Anthropocene as Earth's New Epoch—after 15 Years of Debate," *Nature* 627, no. 8003 (March 6, 2024): 249–50.

10. "Working Group on the 'Anthropocene,'" Subcommission on Quaternary Stratigraphy, accessed January 27, 2024, http://quaternary.stratigraphy.org/working-groups/anthropocene/.

11. "Locating the Anthropocene," Max-Planck-Gesellschaft, July 11, 2023, https://www.mpg.de/20614579/crawford-lake-anthropocene.

12. Ove Hoegh-Guldberg, "Coral Reefs in the Anthropocene: Persistence or the End of the Line?," *Geological Society, London, Special Publications* 395 (May 14, 2014): 167–83.

### Chapter 1. Welcome to the Rock Factory

1. John E. Mylroie and James L. Carew, *Field Guide to the Geology and Karst Geomorphology of San Salvador Island*, 2010, https://geraceresearchcentre.com/pdfs/GeologyKarstSanSal_JMylroieJCarew.pdf; R. Laurence Davis, "Karst Processes and Landforms on San Salvador

Island, Bahamas," Vignettes: Key Concepts in Geomorphology, accessed June 22, 2023, https://serc.carleton.edu/vignettes/collection/43035.html.

2. "Corals: NOAA's National Ocean Service Education."

3. Dennis K. Hubbard, Arnold I. Miller, and David Scaturo, "Production and Cycling of Calcium Carbonate in a Shelf-Edge Reef System (St. Croix, U.S. Virgin Islands): Applications to the Nature of Reef Systems in the Fossil Record," *Journal of Sedimentary Research* 60, no. 3 (May 1, 1990): 335–60.

4. Carolina Castro-Sanguino, Yves-Marie Bozec, and Peter J. Mumby, "Dynamics of Carbonate Sediment Production by *Halimeda*: Implications for Reef Carbonate Budgets," *Marine Ecology Progress Series* 639 (April 2, 2020): 91–106; H. Gray Multer and Ileana E. Clavijo, "Halimeda Investigations: Progress and Problems," in *12th Caribbean Geological Conference* (St. Croix, US Virgin Islands, 1989), 116–27.

5. Mara R. Diaz et al., "Geochemical Evidence of Microbial Activity within Ooids," *Sedimentology* 62, no. 7 (2015): 2090–112.

6. Ulrike Brehm, Wolfgang E. Krumbein, and Katarzyna A. Palinska, "Biomicrospheres Generate Ooids in the Laboratory," *Geomicrobiology Journal* 23, no. 7 (October 1, 2006): 545–50.

7. Diaz et al., "Geochemical Evidence."

8. Rachel Wood, *Reef Evolution* (Oxford: Oxford University Press, 1999), 277.

9. Chris T. Perry et al., "Fish as Major Carbonate Mud Producers and Missing Components of the Tropical Carbonate Factory," *Proceedings of the National Academy of Sciences* 108, no. 10 (March 8, 2011): 3865–69.

10. Perry et al.

11. Anthony J. Martin, *Life Sculpted: Tales of the Animals, Plants, and Fungi That Drill, Break, and Scrape to Shape the Earth* (Chicago: University of Chicago Press, 2023), 65.

12. Wood, *Reef Evolution*.

13. Vincent Caron et al., "Demise and Recovery of Antillean Shallow Marine Carbonate Factories Adjacent to Active Submarine Volcanoes (Lutetian-Bartonian Limestones, St. Bartholomew, French West Indies)," *Sedimentary Geology* 387 (June 1, 2019): 104–25.

14. Wood, *Reef Evolution*.

15. A. Ganopolski, R. Winkelmann, and H. J. Schellnhuber, "Critical Insolation–$CO_2$ Relation for Diagnosing Past and Future Glacial Inception," *Nature* 529, no. 7585 (January 2016): 200–203.

16. William J. Broad, "How the Ice Age Shaped New York," *New York Times*, June 5, 2018, https://www.nytimes.com/2018/06/05/science/how-the-ice-age-shaped-new-york.html.

17. Brian White et al., "Shallowing-Upward Sequence in a Pleistocene Coral Reef and Associated Facies, San Salvador, Bahamas," *AAPG Bulletin* 68, no. 4 (April 1, 1984): 539.

18. William F. Ruddiman, *Earth's Climate: Past and Future* (New York: Macmillan, 2001); George J. Kukla et al., "Last Interglacial Climates," *Quaternary Research* 58, no. 1 (July 2002): 2–13; J. H. Chen et al., "Precise Chronology of the Last Interglacial Period: $^{234}$U-$^{230}$Th Data from Fossil Coral Reefs in the Bahamas," *GSA Bulletin* 103, no. 1 (January 1, 1991): 82–97.

19. White et al., "Shallowing-Upward Sequence"; H. Allen Curran and Brian White, "The Cockburn Town Fossil Coral Reef of San Salvador Island, Bahamas," in *Pleistocene and Holocene*

*Carbonate Environments on San Salvador Island, Bahamas: San Salvador Island, Bahamas, July 2–7, 1989*, vol. 175, ed. H. Allen Curran et al. (Washington, DC: American Geophysical Union, 1989), 27–34.

20. Jeremy B. C. Jackson, "Pleistocene Perspectives on Coral Reef Community Structure," *American Zoologist* 32, no. 6 (December 1, 1992): 719–31.

21. Adam Skarke, "Formation of the Florida Escarpment," NOAA Ocean Exploration, accessed June 5, 2023, https://oceanexplorer.noaa.gov/okeanos/explorations/ex1803/logs/may1/welcome.html.

## Chapter 2. The Present as the Key to the Past

1. Sally E. Walker, Steven M. Holland, and Lisa Gardiner, "*Coenobichnus currani* (New Ichnogenus and Ichnospecies): Fossil Trackway of a Land Hermit Crab, Early Holocene, San Salvador, Bahamas," *Journal of Paleontology* 77, no. 3 (May 2003): 576–82.

2. Jack Repcheck, *The Man Who Found Time* (New York: Perseus, 2003), 41–43.

3. "Geological Pioneers: James Hutton (1726–1797)," Edinburgh Geological Society, accessed June 7, 2023, https://www.edinburghgeolsoc.org/edinburghs-geology/geological-pioneers/james-hutton/.

4. Repcheck, *Man Who Found Time*, 114.

5. "Geological Pioneers: James Hutton."

6. James Hutton, "Theory of the Earth; or an Investigation of the Laws Observable in the Composition, Dissolution, and Restoration of Land upon the Globe," *Earth and Environmental Science Transactions of the Royal Society of Edinburgh* 1, no. 2 (January 1788): 209–304.

7. Repcheck, *Man Who Found Time*, 14.

8. *Siccar Point: Hutton's Unconformity* (Lothian and Borders GeoConservation, n.d.), https://edinburghgeolsoc.org/downloads/Siccar-Point-LBGC-leaflet.pdf.

9. Repcheck, *Man Who Found Time*, 13; John Playfair, "Biographical Account of the Late Dr. James Hutton," *Transactions of the Royal Society of Edinburgh* 5, part 3 (1805): 39–100.

10. Playfair, "Biographical Account."

11. Repcheck, *Man Who Found Time*, 19–21.

12. *Siccar Point: Hutton's Unconformity*.

13. Hutton, "Theory of the Earth."

14. Repcheck, *Man Who Found Time*, 15.

15. Playfair, "Biographical Account."

16. Playfair.

17. Repcheck, *Man Who Found Time*, 161.

18. Repcheck, 169–70.

19. Charles Lyell, *Principles of Geology: Being an Attempt to Explain the Former Changes of the Earth's Surface by Reference to Causes Now in Operation*, 3 vols. (London: John Murray, 1830–1833).

20. Darlene Richardson, "Women in Science: Rediscovering the Accomplishments of Women," in *Women's Work: A Survey of Scholarship by and about Women*, ed. Donna M. Ashcraft (New York: Harrington Park Press, 1998), 45–60.

21. Lyell, *Principles of Geology*.

22. Alistair Sponsel, *Darwin's Evolving Identity: Adventure, Ambition, and the Sin of Speculation* (Chicago: University of Chicago Press, 2018).

23. Sponsel, 43, 69.

24. Charles Darwin, *Charles Darwin's Letters: A Selection, 1825–1859* (Cambridge: Cambridge University Press, 1996).

25. Lyell, *Principles of Geology*.

26. Alistair Sponsel, "Lords of the Ring: Beneath the Surface of the Atoll," *Cabinet Magazine*, 2010, https://cabinetmagazine.org/issues/38/sponsel.php.

27. Charles Darwin, *The Voyage of the* Beagle, Harvard Classics (New York: P. F. Collier & Son, 1909).

28. Darwin, 314.

29. Darwin, 314.

30. Darwin, 314.

31. Charles Darwin, *The Structure and Distribution of Coral Reefs* (Boston: D. Appleton, 1897), 128.

32. Sponsel, *Darwin's Evolving Identity*, 65–67.

33. Darwin, *Voyage of the* Beagle, 410.

34. Sponsel, *Darwin's Evolving Identity*, 70.

35. Darwin, *Structure and Distribution of Coral Reefs*, 95–108.

36. Darwin.

37. Darwin, *Voyage of the* Beagle, 485.

38. Darwin, 469.

39. Darwin, *Structure and Distribution of Coral Reefs*, 21.

40. Darwin, *Voyage of the* Beagle; Darwin, *Structure and Distribution of Coral Reefs*.

41. "History, Culture and Language," Cocos Keeling Islands, accessed June 7, 2023, https://cocoskeelingislands.com.au/history.

42. Darwin, *Voyage of the* Beagle, 456.

43. "History, Culture and Language."

44. Darwin, *Voyage of the* Beagle, 456.

45. "Cocos (Keeling) Islands," NASA Earth Observatory, December 27, 2011, https://earthobservatory.nasa.gov/images/76791/cocos-keeling-islands.

46. Darwin, *Voyage of the* Beagle, 469.

47. Darwin, 471.

48. Darwin, 471–72.

49. Darwin, *Structure and Distribution of Coral Reefs*, 30.

50. Sponsel, "Lords of the Ring."

51. Sponsel, 71–72.

52. Darwin, *Structure and Distribution of Coral Reefs*, 103.

53. Darwin, *Voyage of the* Beagle.

54. David Dobbs, *Reef Madness: Charles Darwin, Alexander Agassiz, and the Meaning of Coral* (New York: Pantheon, 2005); Sponsel, *Darwin's Evolving Identity*.

55. Sponsel, *Darwin's Evolving Identity*, 118.

56. Dobbs, *Reef Madness*.

57. Sponsel, *Darwin's Evolving Identity*, 122.

58. Charles Lyell to Charles Darwin, February 13, 1837, Darwin Correspondence Project, accessed June 7, 2023, https://www.darwinproject.ac.uk/letter/?docId=letters/DCP-LETT-343.xml.

59. Darwin, *Voyage of the* Beagle, 472.

60. Dobbs, *Reef Madness*.

61. Sponsel, *Darwin's Evolving Identity*.

62. John Stanley Gardiner, *Coral Reefs and Atolls* (New York: Macmillan, 1931).

63. Gardiner, 162.

64. H. S. Ladd and S. O. Schlanger, "Bikini and Nearby Atolls, Marshall Islands; Drilling Operations on Eniwetok Atoll," USGS Professional Paper 260-Y, 1960.

65. Ladd and Schlanger.

66. Ladd and Schlanger; Sponsel, "Lords of the Ring."

67. Reginald Aldworth Daly, "The Glacial-Control Theory of Coral Reefs," in *Proceedings of the American Academy of Arts and Sciences* 51 (1915): 158–248.

68. Michael Toomey, Andrew D. Ashton, and J. Taylor Perron, "Profiles of Ocean Island Coral Reefs Controlled by Sea-Level History and Carbonate Accumulation Rates," *Geology* 41, no. 7 (July 1, 2013): 731–34.

69. C. D. Woodroffe et al., "Last Interglacial Reef and Subsidence of the Cocos (Keeling) Islands, Indian Ocean," *Marine Geology* 96, no. 1 (January 1, 1991): 137–43.

70. Darwin, *Structure and Distribution of Coral Reefs*, 174.

## Chapter 3. Reefs at the Shallow End of Deep Time

1. Victoria Egerton and John Mylroie, "Preliminary Investigations of 13 Fossil Reefs on San Salvador Island, Bahamas," in *Proceedings of the 12th Symposium on the Geology of the Bahamas and Other Carbonate Regions*, ed. R. Laurence Davis and Douglas W. Gamble (San Salvador, Bahamas: Gerace Research Centre, 2006), 13–23.

2. Brian White, "Field Guide to the Sue Point Fossil Coral Reefs, San Salvador Island, Bahamas," in *Proceedings of the 4th Symposium on the Geology of the Bahamas and Other Carbonate Regions* (San Salvador, Bahamas: Bahamian Field Station, 1988), 353–65.

3. White et al., "Shallowing-Upward Sequence"; H. Allen Curran and Brian White, "Field Guide to the Cockburn Town Fossil Reef, San Salvador, Bahamas," in *Proceedings of the 2nd Symposium on the Geology of the Bahamas*, ed. James W. Teeter (San Salvador, Bahamas: CCFL Bahamian Field Station, 1984), 71–96.

4. Lisa Gardiner, "Stability of Late Pleistocene Reef Mollusks from San Salvador Island, Bahamas," *PALAIOS* 16, no. 4 (August 1, 2001): 372–86.

5. White et al., "Shallowing-Upward Sequence"; Curran and White, "Field Guide to the Cockburn Town Fossil Reef."

6. Mylroie and Carew, *Field Guide to the Geology*.

7. M. C. Pace, J. E. Mylroie, and J. L. Carew, "Investigation and Review of Dissolution Features on San Salvador Island, Bahamas," in *Proceedings of the 6th Symposium on the Geology of the Bahamas*, ed. Brian White (San Salvador, Bahamas: Bahamian Field Station, 1993), 109–23.

8. Mylroie and Carew, *Field Guide to the Geology*.

9. Curran and White, "Field Guide to the Cockburn Town Fossil Reef."

10. John M. Pandolfi, "The Paleoecology of Coral Reefs," in *Coral Reefs: An Ecosystem in Transition*, ed. Zvy Dubinsky and Noga Stambler (Dordrecht: Springer Netherlands, 2011), 13–24.

11. Curran and White, "Field Guide to the Cockburn Town Fossil Reef."

12. Alexandra Skrivanek, Jin Li, and Andrea Dutton, "Relative Sea-Level Change during the Last Interglacial as Recorded in Bahamian Fossil Reefs," *Quaternary Science Reviews* 200 (November 15, 2018): 160–77.

13. Classic descriptions of reef zones can be found in the following: Goreau, "The Ecology of Jamaican Coral Reefs I. Species Composition and Zonation"; Goreau, Goreau, and Goreau, "Corals and Coral Reefs"; M. A. Huston, "Patterns of Species Diversity on Coral Reefs," *Annual Review of Ecology and Systematics* 16, no. 1 (1985): 149–77.

14. "Elkhorn Coral," NOAA Fisheries, March 8, 2023, https://www.fisheries.noaa.gov/species/elkhorn-coral.

15. "Elkhorn Coral."

16. Huston, "Patterns of Species Diversity."

17. Huston.

18. Jörn Geister, "Calm-Water Reefs and Rough-Water Reefs of the Caribbean Pleistocene," *Acta Palaeontologica Polonica* 25 (1980): 3–4.

19. Geister.

20. Ari Daniel, "Protecting the World's Vanishing Coral Reefs," *Slice of MIT*, April 16, 2021, https://alum.mit.edu/slice/protecting-worlds-vanishing-coral-reefs.

21. John W. Wells, "Memorial to Thomas Fritz Goreau, 1924–1970," Geological Society of America, 1970, https://rock.geosociety.org/net/documents/gsa/memorials/v02/Goreau-TF.pdf.

22. Tom Goreau, "In Memoriam: Dr. Nora Goreau April 25 1921–December 18 2016—Mother of Coral Reef Science," *Global Coral Reef Alliance* (blog), January 15, 2017, https://www.globalcoral.org/memoriam-dr-nora-goreau-april-25-1921-december-18-2016/.

23. Goreau.

24. Goreau; Wells, "Memorial to Thomas Fritz Goreau."

25. Goreau, "Ecology of Jamaican Coral Reefs."

26. Curran and White, "Field Guide to the Cockburn Town Fossil Reef."

27. Peter U. Clark et al., "Oceanic Forcing of Penultimate Deglacial and Last Interglacial Sea-Level Rise," *Nature* 577, no. 7792 (January 2020): 660–64.

28. Brian White and H. Allen Curran, "Coral Reef to Eolianite Transition in the Pleistocene Rocks of Great Inagua, Bahamas," in *Proceedings of the Third Symposium on the Geology of the Bahamas*, ed. H. Allen Curran (San Salvador, Bahamas: Bahamian Field Station, 1987), 165–79; Skrivanek, Li, and Dutton, "Relative Sea-Level Change"; Brian White and H. Allen Curran, "Entombment and Preservation of Sangamonian Coral Reefs during Glacioeustatic Sea-Level Fall, Great Inagua Island, Bahamas," in *Terrestrial and Shallow Marine Geology of the Bahamas and Bermuda*, Geological Society of America Special Paper 300 (Boulder, CO: Geological Society of America, 1995), 51–62.

29. Richard E. Dodge et al., "Pleistocene Sea Levels from Raised Coral Reefs of Haiti," *Science* 219, no. 4591 (March 25, 1983): 1423–25; Patrick Schielein et al., "ESR and U-Th Dating

Results for Last Interglacial Coral Reef Terraces at the Northern Coast of Cuba," *Quaternary International*, 556 (August 10, 2020): 216–29.

30. William F. Precht and Richard Aronson, "Death and Resurrection of Caribbean Coral Reefs: A Palaeoecological Perspective," in *Coral Reef Conservation*, ed. Isabelle Côté and John Reynolds (Cambridge: Cambridge University Press, 2006), 40–77.

31. William F. Precht and Steven L. Miller, "Ecological Shifts along the Florida Reef Tract: The Past as a Key to the Future," in *Geological Approaches to Coral Reef Ecology*, ed. Richard B. Aronson (New York: Springer, 2007), 237–312; B. J. Greenstein, H. A. Curran, and J. M. Pandolfi, "Shifting Ecological Baselines and the Demise of *Acropora cervicornis* in the Western North Atlantic and Caribbean Province: A Pleistocene Perspective," *Coral Reefs* 17, no. 3 (September 1, 1998): 249–61; E. Gischler, J. H. Hudson, and D. Storz, "Growth of Pleistocene Massive Corals in South Florida: Low Skeletal Extension-Rates and Possible ENSO, Decadal, and Multi-Decadal Cyclicities," *Coral Reefs* 28, no. 4 (December 1, 2009): 823–30.

32. Robert L. Folk, Harry H. Roberts, and Clyde H. Moore, "Black Phytokarst from Hell, Cayman Islands, British West Indies," *GSA Bulletin* 84, no. 7 (July 1, 1973): 2351–60.

33. B. Jones and I. G. Hunter, "Pleistocene Paleogeography and Sea Levels on the Cayman Islands, British West Indies," *Coral Reefs* 9, no. 2 (September 1, 1990): 81–91; I. G. Hunter and B. Jones, "Coral Associations of the Pleistocene Ironshore Formation, Grand Cayman," *Coral Reefs* 15, no. 4 (November 1, 1996): 249–67.

34. Paul Blanchon et al., "Rapid Sea-Level Rise and Reef Back-Stepping at the Close of the Last Interglacial Highstand," *Nature* 458, no. 7240 (April 2009): 881–84; Paul Blanchon, "Reef Demise and Back-Stepping during the Last Interglacial, Northeast Yucatan," *Coral Reefs* 29, no. 2 (June 1, 2010): 481–98.

35. Blanchon, "Reef Demise and Back-Stepping during the Last Interglacial."

36. Ulrich Radtke and Gerhard Schellmann, "Uplift History along the Clermont Nose Traverse on the West Coast of Barbados during the Last 500,000 Years—Implications for Paleo-Sea Level Reconstructions," *Journal of Coastal Research* 22, no. 2 (222) (March 1, 2006): 350–56.

37. John M. Pandolfi and Jeremy B. C. Jackson, "Ecological Persistence Interrupted in Caribbean Coral Reefs," *Ecology Letters* 9, no. 7 (2006): 818–26.

38. Pandolfi and Jackson.

39. J. M. Pandolfi, G. Llewellyn, and J. B. C. Jackson, "Pleistocene Reef Environments, Constituent Grains, and Coral Community Structure: Curaçao, Netherlands Antilles," *Coral Reefs* 18, no. 2 (July 1, 1999): 107–22.

40. John M. Pandolfi and Jeremy B. C. Jackson, "Community Structure of Pleistocene Coral Reefs of Curaçao, Netherlands Antilles," *Ecological Monographs* 71, no. 1 (2001): 49–67; Pandolfi, Llewellyn, and Jackson, "Pleistocene Reef Environments."

41. Karen Vyverberg et al., "Episodic Reef Growth in the Granitic Seychelles during the Last Interglacial: Implications for Polar Ice Sheet Dynamics," *Marine Geology* 399 (May 1, 2018): 170–87; Andrea Dutton et al., "Tropical Tales of Polar Ice: Evidence of Last Interglacial Polar Ice Sheet Retreat Recorded by Fossil Reefs of the Granitic Seychelles Islands," *Quaternary Science Reviews* 107 (January 1, 2015): 182–96.

42. Evan N. Edinger, John M. Pandolfi, and Russell A. Kelley, "Community Structure of Quaternary Coral Reefs Compared with Recent Life and Death Assemblages," *Paleobiology* 27, no. 4 (2001): 669–94; John M. Pandolfi, "Limited Membership in Pleistocene Reef Coral

Assemblages from the Huon Peninsula, Papua New Guinea: Constancy during Global Change," *Paleobiology* 22, no. 2 (April 1996): 152–76.

43. Pandolfi, "Limited Membership."

44. Pandolfi.

45. Lorraine R. Casazza, "Pleistocene Reefs of the Egyptian Red Sea: Environmental Change and Community Persistence," *PeerJ* 5 (June 28, 2017): e3504.

46. Casazza.

47. Robert C. Walter et al., "Early Human Occupation of the Red Sea Coast of Eritrea during the Last Interglacial," *Nature* 405, no. 6782 (May 2000): 65–69.

48. Walter et al.

49. Goreau, Goreau, and Goreau, "Corals and Coral Reefs."

50. J. D. Woodley et al., "Hurricane Allen's Impact on Jamaican Coral Reefs," *Science* 214, no. 4522 (1981): 749–55.

51. Jeremy Jackson et al., eds., *Status and Trends of Caribbean Coral Reefs: 1970–2012* (Gland, Switzerland: Global Coral Reef Monitoring Network; International Union for the Conservation of Nature, 2014), https://bvearmb.do/handle/123456789/590.

52. Nancy Knowlton, "The Future of Coral Reefs," *Proceedings of the National Academy of Sciences* 98, no. 10 (May 8, 2001): 5419–25.

53. Wendy T. Muraoka et al., "Historical Declines in Parrotfish on Belizean Coral Reefs Linked to Shifts in Reef Exploitation Following European Colonization," *Frontiers in Ecology and Evolution* 10 (2022).

54. Thomas J. Goreau, "Bleaching and Reef Community Change in Jamaica: 1951–1991," *American Zoologist* 32 (1992): 683–95.

55. *Coral Ghosts*, CBC Documentary Channel, accessed June 6, 2023, https://www.cbc.ca/documentarychannel/docs/coral-ghosts.

56. *Coral Ghosts*.

57. *Coral Ghosts*.

58. Precht and Aronson, "Death and Resurrection."

59. George Roff and Peter J. Mumby, "Global Disparity in the Resilience of Coral Reefs," *Trends in Ecology & Evolution* 27, no. 7 (July 1, 2012): 404–13.

60. Roff and Mumby.

61. A. F. Budd, "Diversity and Extinction in the Cenozoic History of Caribbean Reefs," *Coral Reefs* 19, no. 1 (April 1, 2000): 25–35.

62. Jeremy B. C. Jackson and Aaron O'Dea, "Evolution and Environment of Caribbean Coastal Ecosystems," *Proceedings of the National Academy of Sciences* 120, no. 42 (October 17, 2023): e2307520120.

63. John M. Pandolfi and Ann F. Budd, "A Festschrift for Jeremy B. C. Jackson and His Integration of Paleobiology, Ecology, Evolution, and Conservation Biology," *Evolutionary Ecology* 26, no. 2 (2012): 227–32.

64. Budd, "Diversity and Extinction."

65. Héctor Reyes-Bonilla and Eric Jordán-Dahlgren, "Caribbean Coral Reefs: Past, Present, and Insights into the Future," in *Marine Animal Forests: The Ecology of Benthic Biodiversity Hotspots*, ed. Sergio Rossi et al. (Cham, Switzerland: Springer, 2017), 31–72.

66. L. W. Smith, *Indo-Pacific Reef-Building Corals: General Status Assessment* (Honolulu: National Marine Fisheries Service, Pacific Islands Regional Office, 2019).

67. Jackson et al., *Status and Trends*.

68. Christopher R. Biggs et al., "Does Functional Redundancy Affect Ecological Stability and Resilience? A Review and Meta-Analysis," *Ecosphere* 11, no. 7 (2020): e03184.

69. T. P. Hughes et al., "Climate Change, Human Impacts, and the Resilience of Coral Reefs," *Science* 301, no. 5635 (August 15, 2003): 929–33.

70. Jennifer K. McWhorter et al., "Climate Refugia on the Great Barrier Reef Fail When Global Warming Exceeds 3°C," *Global Change Biology* 28, no. 19 (2022): 5768–80.

71. Tara R. Clark et al., "U-Th Dating Reveals Regional-Scale Decline of Branching *Acropora* Corals on the Great Barrier Reef over the Past Century," *Proceedings of the National Academy of Sciences* 114, no. 39 (September 26, 2017): 10350–55.

72. Pandolfi and Jackson, "Ecological Persistence"; Jackson, "Pleistocene Perspectives"; Greenstein, Curran, and Pandolfi, "Shifting Ecological Baselines."

73. John M. Pandolfi and Jeremy B. C. Jackson, "Broad-Scale Patterns in Pleistocene Coral Reef Communities from the Caribbean: Implications for Ecology and Management," in *Geological Approaches to Coral Reef Ecology*, ed. Richard B. Aronson (New York: Springer, 2007), 201–36.

74. Pandolfi and Jackson.

75. H. Allen Curran et al., "The Health and Short-Term Change of Two Coral Patch Reefs, Fernandez Bay, San Salvador Island, Bahamas," *Oceanographic Literature Review* 8, no. 42 (1995): 674–75; H. Allen Curran et al., "Shallow-Water Reefs in Transition: Examples from Belize and the Bahamas," in *Proceedings of the 11th Symposium on the Geology of the Bahamas and Other Carbonate Regions*, ed. Ronald D. Lewis and Bruce C. Panuska (San Salvador, Bahamas: Gerace Research Center, 2003), 13–24.

76. Curran et al., "Health and Short-Term Change"; Curran et al., "Shallow-Water Reefs in Transition."

77. Curran et al., "Health and Short-Term Change"; Curran et al., "Shallow-Water Reefs in Transition."

78. Paulette Peckol et al., "Assessment of Coral Reefs off San Salvador Island, Bahamas," *Atoll Research Bulletin* 496, no. 7 (2003): 124.

79. Curran et al., "Shallow-Water Reefs in Transition."

80. Jackson et al., *Status and Trends*.

81. Greenstein, Curran, and Pandolfi, "Shifting Ecological Baselines."

## Chapter 4. Into the Death Assemblage

1. "Florida Reefs," GulfBase, Texas A&M University, accessed June 7, 2023, https://www.gulfbase.org/geological-feature/florida-reefs.

2. "Florida," Coral Reef Information System, National Oceanic and Atmospheric Administration, accessed June 7, 2023, https://www.coris.noaa.gov/portals/florida.html.

3. "Florida Reefs."

4. J. Candace Clifford, ed., *Inventory of Historic Light Stations* (Washington, DC: National Maritime Initiative, National Park Service, History Division, 1994).

5. The US government sold the lighthouse in 2022. It's not known what it may be used for in the future. For information about the sale, see "GSA Announces Public Auction for Three Historic Lighthouses in Florida Keys," US General Services Administration, February 24, 2022, https://www.gsa.gov/about-us/regions/region-4-southeast-sunbelt/region-4-newsroom/news-releases/gsa-announces-public-auction-for-three-historic-lighthouses-in-florida-keys-02242022.

6. Martin, *Life Sculpted*, 57–58.

7. "Carysfort Reef Lighthouse," LighthouseFriends, accessed June 23, 2023, https://www.lighthousefriends.com/light.asp?ID=703.

8. John Pickrell, "How Can I Become a Fossil?," BBC, February 15, 2018, https://www.bbc.com/future/article/20180215-how-does-fossilisation-happen.

9. Anna K. Behrensmeyer, Christiane Denys, and Jean-Philip Brugal, "What Is Taphonomy and What Is Not?," *Historical Biology* 30, no. 6 (August 18, 2018): 718–19.

10. George Grebenschikov, "Efremov's Science Fiction: A Reexamination of His Major Works," *Russian Language Journal / Русский Язык* 30, no. 106 (1976): 105–14.

11. I. A. Efremov, "Taphonomy: A New Branch of Paleontology," *Pan-American Geologist* 74 (1940): 81–93.

12. John E. Warme, "Actualistic Paleontology," in *Paleontology*, Encyclopedia of Earth Sciences Series (Berlin: Springer, 1979), 4–10.

13. Gary E. Strathearn, "*Homotrema rubrum*: Symbiosis Identified by Chemical and Isotopic Analyses," *PALAIOS* 1, no. 1 (1986): 48–54.

14. Hubbard, Miller, and Scaturo, "Production and Cycling of Calcium Carbonate," fig. 5.

15. Benjamin J. Greenstein and John M. Pandolfi, "Taphonomic Alteration of Reef Corals: Effects of Reef Environment and Coral Growth Form II: The Florida Keys," *PALAIOS* 18, no. 6 (2003): 495–509.

16. Greenstein and Pandolfi, "Taphonomic Alteration of Reef Corals."

17. Benjamin J. Greenstein and John M. Pandolfi, "Preservation of Community Structure in Modern Reef Coral Life and Death Assemblages of the Florida Keys: Implications for the Quaternary Fossil Record of Coral Reefs," *Bulletin of Marine Science* 61, no. 2 (September 1, 1997): 431–52.

18. Greenstein and Pandolfi; John M. Pandolfi and Benjamin J. Greenstein, "Preservation of Community Structure in Death Assemblages of Deep-Water Caribbean Reef Corals," *Limnology and Oceanography* 42, no. 7 (1997): 1505–16.

19. Pandolfi and Greenstein, "Preservation of Community Structure in Death Assemblages of Deep-Water Caribbean Reef Corals."

20. Pandolfi and Greenstein.

21. Susan M. Kidwell, "Time-Averaging and Fidelity of Modern Death Assemblages: Building a Taphonomic Foundation for Conservation Palaeobiology," *Palaeontology* 56, no. 3 (2013): 487–522.

22. Greenstein, Curran, and Pandolfi, "Shifting Ecological Baselines."

23. Clark et al., "U-Th Dating Reveals Regional-Scale Decline."

24. Lisa Greer et al., "How Vulnerable Is *Acropora cervicornis* to Environmental Change? Lessons from the Early to Middle Holocene," *Geology* 37, no. 3 (March 1, 2009): 263–66.

25. Greer et al.
26. Aronson and Precht, "Stasis, Biological Disturbance."
27. Aronson and Precht.
28. Aronson and Precht.
29. Aronson and Precht.
30. Greenstein, Curran, and Pandolfi, "Shifting Ecological Baselines."
31. Jackson et al., *Status and Trends*.
32. "US Coral Reef Monitoring Data Summary 2018," NOAA Technical Memorandum (NOAA Coral Reef Conservation Program, 2018); *Coral Reef Condition: A Status Report for Florida's Coral Reef* (NOAA Coral Reef Conservation Program, 2020).
33. Jackson et al., *Status and Trends*, 99.
34. "Extreme Ocean Temperatures Are Affecting Florida's Coral Reef," National Environmental Satellite, Data, and Information Service, August 18, 2023, https://www.nesdis.noaa.gov/news/extreme-ocean-temperatures-are-affecting-floridas-coral-reef.
35. *Coral Reef Condition*.
36. "Ocean Acidification May Be Impacting Coral Reefs in the Florida Keys," American Geophysical Union Newsroom, accessed June 7, 2023, https://news.agu.org/press-release/ocean-acidification-may-be-impacting-coral-reefs-in-the-florida-keys/.

## Chapter 5. When Reefs Fall Apart

1. Jackson et al., *Status and Trends*.
2. James Crabbe et al., "*Acropora cervicornis*," IUCN Red List of Threatened Species, June 1, 2021, https://www.iucnredlist.org/en; James Crabbe et al., "*Acropora palmata*," IUCN Red List of Threatened Species, June 1, 2021, https://www.iucnredlist.org/en.
3. "IUCN Red List of Threatened Species," accessed June 9, 2023, https://www.iucnredlist.org/en.
4. Jackson et al., *Status and Trends*.
5. Glenn De'ath et al., "The 27-Year Decline of Coral Cover on the Great Barrier Reef and Its Causes," *Proceedings of the National Academy of Sciences* 109, no. 44 (October 30, 2012): 17995–99; Jeremy B. C. Jackson et al., "Historical Overfishing and the Recent Collapse of Coastal Ecosystems," *Science* 293, no. 5530 (July 27, 2001): 629–37.
6. Jackson et al., *Status and Trends*; Toby A. Gardner et al., "Long-Term Region-Wide Declines in Caribbean Corals," *Science* 301, no. 5635 (August 15, 2003): 958–60; Goreau, "Bleaching and Reef Community Change."
7. Jackson et al., *Status and Trends*.
8. Michael Marshall, "Bacterial Suspects Identified in Caribbean Coral Deaths," *New Scientist*, June 18, 2014, https://www.newscientist.com/article/dn25743-bacterial-suspects-identified-in-caribbean-coral-deaths/.
9. Ed Yong, *I Contain Multitudes: The Microbes within Us and a Grander View of Life* (New York: Ecco, 2016).
10. Jackson et al., *Status and Trends*.
11. Sean Cummings, "Mysterious Sea Urchin Plague Is Spreading through the World's Oceans," American Association for the Advancement of Science, accessed June 4, 2024,

https://www.science.org/content/article/mysterious-sea-urchin-plague-spreading-through-world-s-oceans.

12. Stephanie M. Rosales et al., "Microbiome Differences in Disease-Resistant vs. Susceptible *Acropora* Corals Subjected to Disease Challenge Assays," *Scientific Reports* 9, no. 1 (December 4, 2019): 18279.

13. Roff and Mumby, "Global Disparity."

14. Jackson et al., *Status and Trends*.

15. Jackson et al., 20.

16. Toby Gardner et al., "Hurricanes and Caribbean Coral Reefs: Impacts, Recovery Patterns, and Role in Long-Term Decline," *Ecology* 86 (January 1, 2005): 174–84.

17. Woodley et al., "Hurricane Allen's Impact."

18. Woodley et al.

19. Randy Olson, "Evolution of a Public Intellectual: Coral Reef Biologist Jeremy Jackson," *Journal of Science Communication* 16, no. 1 (March 28, 2017): C04.

20. Woodley et al., "Hurricane Allen's Impact."

21. Precht and Aronson, "Death and Resurrection."

22. Jeremy Jackson, "How We Wrecked the Ocean," TED Talk, April 2010, https://www.ted.com/talks.jeremy_jackson_how_we_wrecked_the_ocean.

23. Katie L. Cramer et al., "The Transformation of Caribbean Coral Communities since Humans," *Ecology and Evolution* 11, no. 15 (2021): 10098–118.

24. Jackson et al., *Status and Trends*.

25. Steve R. Dudgeon et al., "Phase Shifts and Stable States on Coral Reefs," *Marine Ecology Progress Series* 413 (2010): 201–16; Tak Fung, Robert M. Seymour, and Craig R. Johnson, "Alternative Stable States and Phase Shifts in Coral Reefs under Anthropogenic Stress," *Ecology* 92, no. 4 (2011): 967–82.

26. Dudgeon et al., "Phase Shifts and Stable States."

27. Pandolfi and Jackson, "Ecological Persistence."

28. Jackson et al., *Status and Trends*.

29. Jackson et al., 14.

30. Dudgeon et al., "Phase Shifts and Stable States."

31. Joshua A. Idjadi et al., "Rapid Phase-Shift Reversal on a Jamaican Coral Reef," *Coral Reefs* 25, no. 2 (May 1, 2006): 209–11.

32. Dudgeon et al., "Phase Shifts and Stable States"; Precht and Aronson, "Death and Resurrection."

33. Precht and Aronson, "Death and Resurrection."

34. Roff and Mumby, "Global Disparity."

35. Jackson et al., "Historical Overfishing."

36. De'ath et al., "27-Year Decline of Coral Cover."

37. Roff and Mumby, "Global Disparity"; De'ath et al., "27-Year Decline of Coral Cover."

38. De'ath et al., "27-Year Decline of Coral Cover."

39. De'ath et al.

40. *Annual Summary Report of Coral Reef Condition 2021/22*, Australian Institute of Marine Science, accessed June 8, 2023, https://www.aims.gov.au/monitoring-great-barrier-reef/gbr-condition-summary-2021-22.

41. "Highest Coral Cover in Central, Northern Reef in 36 Years," Australian Institute of Marine Science, August 4, 2022, https://www.aims.gov.au/information-centre/news-and-stories/highest-coral-cover-central-northern-reef-36-years.

42. Jeremy B. C. Jackson and Karen E. Alexander, "Introduction: The Importance of Shifting Baselines," in *Shifting Baselines: The Past and the Future of Ocean Fisheries*, ed. Jeremy B. C. Jackson, Karen E. Alexander, and Enric Sala (Washington, DC: Island Press, 2011), 1–8.

43. Jackson and Alexander; Daniel Pauly, "Anecdotes and the Shifting Baseline Syndrome of Fisheries," *Trends in Ecology & Evolution* 10, no. 10 (October 1995).

44. Knowlton and Jackson, "Shifting Baselines."

45. Roy Waldo Miner, "Diving in Coral Gardens," *Scientific American* 151, no. 3 (September 1934): 122–24.

46. "Andros Coral Reef Diorama," American Museum of Natural History, accessed June 8, 2023, https://www.amnh.org/exhibitions/permanent/ocean-life/andros-coral-reef-diorama.

47. Miner, "Diving in Coral Gardens."

48. J. B. C. Jackson, "Reefs since Columbus," *Coral Reefs* 16, no. 1 (June 1, 1997): S23–32.

49. Loren McClenachan et al., "Ghost Reefs: Nautical Charts Document Large Spatial Scale of Coral Reef Loss over 240 Years," *Science Advances* 3, no. 9 (September 6, 2017): e1603155.

50. McClenachan et al.

51. McClenachan et al.

52. Cramer et al., "Transformation of Caribbean Coral Communities."

53. Cramer et al.

54. Katie L. Cramer et al., "Widespread Loss of Caribbean Acroporid Corals Was Underway before Coral Bleaching and Disease Outbreaks," *Science Advances* 6, no. 17 (April 22, 2020): eaax9395.

55. Cramer et al., "Transformation of Caribbean Coral Communities."

56. Darwin, *Voyage of the* Beagle.

57. McClenachan et al., "Ghost Reefs."

58. Peter Chapman, *Bananas: How the United Fruit Company Shaped the World* (Edinburgh, UK: Canongate Books, 2022).

59. Katie L. Cramer, "History of Human Occupation and Environmental Change in Western and Central Caribbean Panama," *Bulletin of Marine Science* 89, no. 4 (October 2013): 955–82.

60. Cramer.

61. Jessica Haapkylä et al., "Seasonal Rainfall and Runoff Promote Coral Disease on an Inshore Reef," *PLOS ONE* 6, no. 2 (February 10, 2011): e16893; Cramer et al., "Transformation of Caribbean Coral Communities."

62. Rebecca L. Vega Thurber et al., "Chronic Nutrient Enrichment Increases Prevalence and Severity of Coral Disease and Bleaching," *Global Change Biology* 20, no. 2 (2014): 544–54.

63. Pandolfi and Jackson, "Ecological Persistence."

64. Nicholas M. Hammerman et al., "Unraveling Moreton Bay Reef History: An Urban High-Latitude Setting for Coral Development," *Frontiers in Ecology and Evolution* 10 (2022).

65. Hammerman et al.

66. Clark et al., "U-Th Dating Reveals Regional-Scale Decline"; George Roff et al., "Palaeoecological Evidence of a Historical Collapse of Corals at Pelorus Island, Inshore Great Barrier

Reef, Following European Settlement," *Proceedings of the Royal Society B: Biological Sciences* 280, no. 1750 (January 7, 2013): 20122100.

67. Jackson, "Reefs since Columbus"; Knowlton and Jackson, "Shifting Baselines."

68. Loren McClenachan, "Documenting Loss of Large Trophy Fish from the Florida Keys with Historical Photographs," *Conservation Biology* 23, no. 3 (June 2009): 636–43.

69. McClenachan.

70. McClenachan.

71. Stuart A. Sandin et al., "Baselines and Degradation of Coral Reefs in the Northern Line Islands," *PLOS ONE* 3, no. 2 (February 27, 2008): e1548.

72. Jackson, "Reefs since Columbus."

73. Jackson.

74. Loren McClenachan, Jeremy B. C. Jackson, and Marah J. H. Newman, "Conservation Implications of Historic Sea Turtle Nesting Beach Loss," *Frontiers in Ecology and the Environment* 4, no. 6 (2006): 290–96.

75. McClenachan, Jackson, and Newman.

76. Jackson, "Reefs since Columbus."

77. McClenachan, Jackson, and Newman, "Conservation Implications."

78. McClenachan, Jackson, and Newman.

79. Peter J. Mumby, "Herbivory versus Corallivory: Are Parrotfish Good or Bad for Caribbean Coral Reefs?," *Coral Reefs* 28, no. 3 (September 1, 2009): 683–90.

80. Robert S. Steneck et al., "Managing Recovery Resilience in Coral Reefs against Climate-Induced Bleaching and Hurricanes: A 15 Year Case Study from Bonaire, Dutch Caribbean," *Frontiers in Marine Science* 6 (2019).

81. Steneck et al.

82. Steneck et al.

83. Andrew A. Shantz, Mark C. Ladd, and Deron E. Burkepile, "Overfishing and the Ecological Impacts of Extirpating Large Parrotfish from Caribbean Coral Reefs," *Ecological Monographs* 90, no. 2 (2020): e01403.

84. "Plenty of Fish?," United Nations Framework Convention on Climate Change, accessed August 28, 2024, https://unfccc.int/news/plenty-of-fish.

85. Jackson et al., "Historical Overfishing."

86. Katie L. Cramer et al., "Millennial-Scale Change in the Structure of a Caribbean Reef Ecosystem and the Role of Human and Natural Disturbance," *Ecography* 43, no. 2 (2020): 283–93; Katie L. Cramer et al., "Prehistorical and Historical Declines in Caribbean Coral Reef Accretion Rates Driven by Loss of Parrotfish," *Nature Communications* 8, no. 1 (January 23, 2017): 14160; Muraoka et al., "Historical Declines in Parrotfish."

87. Cramer et al., "Millennial-Scale Change."

88. Cramer et al., "Prehistorical and Historical Declines."

89. Cramer et al., "Prehistorical and Historical Declines."

90. Shantz, Ladd, and Burkepile, "Overfishing and the Ecological Impacts"; L. T. Toth et al., "Do No-Take Reserves Benefit Florida's Corals? 14 Years of Change and Stasis in the Florida Keys National Marine Sanctuary," *Coral Reefs* 33, no. 3 (September 1, 2014): 565–77.

91. Roff and Mumby, "Global Disparity."

92. Joseph H. Connell, "Diversity in Tropical Rain Forests and Coral Reefs," *Science* 199, no. 4335 (March 24, 1978): 1302–10.

93. Connell.

94. Connell.

95. J. M. Pandolfi et al., "Mass Mortality Following Disturbance in Holocene Coral Reefs from Papua New Guinea," *Geology* 34, no. 11 (November 1, 2006): 949–52.

96. Pandolfi et al.

97. Pandolfi et al.

98. J. H. Connell, "Disturbance and Recovery of Coral Assemblages," *Coral Reefs* 16, no. 1 (June 1, 1997): S101–13.

99. "2023 Was the World's Warmest Year on Record, by Far," National Oceanic and Atmospheric Administration, January 12, 2024, https://www.noaa.gov/news/2023-was-worlds-warmest-year-on-record-by-far.

100. "Global $CO_2$ Levels," 2° Institute, accessed June 7, 2023, https://www.co2levels.org/.

101. Rebecca Lindsey, "Climate Change: Atmospheric Carbon Dioxide," NOAA, April 9, 2024, http://www.climate.gov/news-features/understanding-climate/climate-change-atmospheric-carbon-dioxide.

102. Rebecca Albright et al., "Ocean Acidification Compromises Recruitment Success of the Threatened Caribbean Coral *Acropora palmata*," *Proceedings of the National Academy of Sciences* 107, no. 47 (November 23, 2010): 20400–404.

103. Albright et al.

104. Nicolas R. Evensen et al., "Scaling the Effects of Ocean Acidification on Coral Growth and Coral–Coral Competition on Coral Community Recovery," *PeerJ* 9 (July 13, 2021): e11608.

105. Knowlton et al., "Rebuilding Coral Reefs."

106. Christopher E. Cornwall et al., "Global Declines in Coral Reef Calcium Carbonate Production under Ocean Acidification and Warming," *Proceedings of the National Academy of Sciences* 118, no. 21 (May 25, 2021): e2015265118.

107. "Integrating Paleo and Historical Data into Coral Reef Management and Policy." Conservation Paleobiology Network, accessed June 8, 2023, https://conservationpaleorcn.org/coral-reef-working-group/.

## Chapter 6. When Reefs Persist

1. Mark A. Wilson, H. Allen Curran, and Brian White, "Paleontological Evidence of a Brief Global Sea-Level Event during the Last Interglacial," *Lethaia* 31, no. 3 (1998): 241–50; Brian White, H. Allen Curran, and Mark Wilson, "Bahamian Sangamonian Coral Reefs and Sea-Level Change," *Proceedings of the Eighth Symposium on the Geology of the Bahamas and Other Carbonate Regions*, ed. James L. Carew (San Salvador, Bahamas: Bahamian Field Station, 1997), 196–213.

2. Sandra Russell, "The Islands of the Bahamas," accessed June 8, 2023, https://bahamasgeotourism.com.

3. White, Curran, and Wilson, "Bahamian Sangamonian Coral Reefs"; Wilson, Curran, and White, "Paleontological Evidence."

4. Skrivanek, Li, and Dutton, "Relative Sea-Level Change."

5. Wilson, Curran, and White, "Paleontological Evidence"; White, Curran, and Wilson, "Bahamian Sangamonian Coral Reefs."

6. Thomas Olszewski, "Taking Advantage of Time-Averaging," *Paleobiology* 25, no. 2 (1999): 226–38.

7. Wilson, Curran, and White, "Paleontological Evidence"; Skrivanek, Li, and Dutton, "Relative Sea-Level Change."

8. Martin, *Life Sculpted*, 47.

9. Wilson, Curran, and White, "Paleontological Evidence."

10. White, Curran, and Wilson, "Bahamian Sangamonian Coral Reefs."

11. K. Miller and C. Mundy, "Rapid Settlement in Broadcast Spawning Corals: Implications for Larval Dispersal," *Coral Reefs* 22, no. 2 (July 1, 2003): 99–106.

12. J. R. Wilson and P. L. Harrison, "Settlement-Competency Periods of Larvae of Three Species of Scleractinian Corals," *Marine Biology* 131, no. 2 (May 1, 1998): 339–45.

13. Past Interglacials Working Group of PAGES, "Interglacials of the Last 800,000 Years," *Reviews of Geophysics* 54, no. 1 (2016): 162–219.

14. Robert E. Kopp, A. Dutton, and A. E. Carlson, "Centennial- to Millennial-Scale Sea-Level Change during the Holocene and Last Interglacial Periods," *Past Global Changes Magazine* 25, no. 3 (December 2017): 148–49.

15. Kopp, Dutton, and Carlson.

16. Matthew Rahamut and Thomas Stemann, "Signals of Two-Stage Accretion of the Late Pleistocene Coral Reef Deposits in Treasure Beach and Rio Bueno, Jamaica," *Abstracts with Programs—Geological Society of America* 54, no. 5 (2022).

17. Kopp, Dutton, and Carlson, "Centennial- to Millennial-Scale Sea-Level Change."

18. Kopp, Dutton, and Carlson.

19. Kopp, Dutton, and Carlson.

20. A. Dutton et al., "Sea-Level Rise Due to Polar Ice-Sheet Mass Loss during Past Warm Periods," *Science* 349, no. 6244 (July 10, 2015): aaa4019.

21. Rebecca Lindsey, "Climate Change: Global Sea Level," NOAA, accessed June 8, 2023, http://www.climate.gov/news-features/understanding-climate/climate-change-global-sea-level.

22. "The Greenland Ice Sheet Is Close to a Melting Point of No Return," American Geophysical Union Newsroom, accessed July 5, 2023, https://news.agu.org/press-release/the-greenland-ice-sheet-is-close-to-a-melting-point-of-no-return/; "Greenland's Ice Is Melting," UCAR Center for Science Education, accessed June 8, 2023, https://scied.ucar.edu/learning-zone/climate-change-impacts/greenlands-ice-melting.

23. Ian G. Macintyre, "Demise, Regeneration, and Survival of Some Western Atlantic Reefs during the Holocene Transgression," in *Geological Approaches to Coral Reef Ecology*, ed. Richard Aronson (New York: Springer, 2007), 181–200.

24. Paul Blanchon, Alexis Medina-Valmaseda, and Fiona D. Hibbert, "Revised Postglacial Sea-Level Rise and Meltwater Pulses from Barbados," *Open Quaternary* 7, no. 1 (May 10, 2021): 1–12.

25. Kelsey L. Sanborn et al., "New Evidence of Hawaiian Coral Reef Drowning in Response to Meltwater Pulse-1A," *Quaternary Science Reviews* 175 (November 1, 2017): 60–72.

26. Gilbert F. Camoin et al., "Reef Response to Sea-Level and Environmental Changes during the Last Deglaciation: Integrated Ocean Drilling Program Expedition 310, Tahiti Sea Level," *Geology* 40, no. 7 (July 1, 2012): 643–46.

27. Camoin et al., "Reef Response to Sea-Level."

28. Jody M. Webster et al., "Response of the Great Barrier Reef to Sea-Level and Environmental Changes over the Past 30,000 Years," *Nature Geoscience* 11, no. 6 (June 2018): 426–32.

29. "About IODP: History," International Ocean Discovery Program, accessed June 8, 2023, https://www.iodp.org/about-iodp/history.

30. Webster et al., "Response of the Great Barrier Reef."

31. *Lunchbox Science with Associate Professor Jody Webster*, 2020, https://www.youtube.com/watch?v=F-50w0ZoSQA.

32. John M. Pandolfi, Jeremy B. C. Jackson, and J. Geister, "Geologically Sudden Natural Extinction of Two Widespread Late Pleistocene Caribbean Reef Corals," in *Evolutionary Patterns: Growth, Form, and Tempo in the Fossil Record*, ed. Jeremy B. C. Jackson, S. Lidgard, and F. K. McKinney (Chicago: University of Chicago Press, 2001), 120–58; John M. Pandolfi, "Response of Pleistocene Coral Reefs to Environmental Change over Long Temporal Scales," *American Zoologist* 39, no. 1 (February 1, 1999): 113–30.

33. John M. Pandolfi, "A New, Extinct Pleistocene Reef Coral from the *Montastraea 'annularis'* Species Complex," *Journal of Paleontology* 81, no. 3 (May 2007): 472–82; Pandolfi, Jackson, and Geister, "Geologically Sudden Natural Extinction."

34. Lauren T. Toth et al., "A New Record of the Late Pleistocene Coral *Pocillopora palmata* from the Dry Tortugas, Florida Reef Tract, USA," *PALAIOS* 30, no. 12 (December 2015): 827–35; Pandolfi, Jackson, and Geister, "Geologically Sudden Natural Extinction."

35. Lauren T. Toth and Richard B. Aronson, "The 4.2ka Event, ENSO, and Coral Reef Development," *Climate of the Past* 15, no. 1 (January 16, 2019): 105–19; Lauren T. Toth et al., "ENSO Drove 2500-Year Collapse of Eastern Pacific Coral Reefs," *Science* 337, no. 6090 (July 6, 2012): 81–84; Lauren T. Toth et al., "Climatic and Biotic Thresholds of Coral-Reef Shutdown," *Nature Climate Change* 5, no. 4 (April 2015): 369–74.

36. Toth and Aronson, "4.2ka Event, ENSO"; Toth et al., "ENSO Drove 2500-Year Collapse"; Toth et al., "Climatic and Biotic Thresholds."

37. Lauren T. Toth, Ian G. Macintyre, and Richard B. Aronson, "Holocene Reef Development in the Eastern Tropical Pacific," in *Coral Reefs of the Eastern Tropical Pacific: Persistence and Loss in a Dynamic Environment*, ed. Peter W. Glynn, Derek P. Manzello, and Ian C. Enochs (Dordrecht: Springer Netherlands, 2017), 177–201.

38. Paul C. Fiedler and Miguel F. Lavín, "Oceanographic Conditions of the Eastern Tropical Pacific," in Glynn, Manzello, and Enochs, *Coral Reefs of the Eastern Tropical Pacific*, 59–83.

39. Toth and Aronson, "4.2ka Event, ENSO"; Toth et al., "ENSO Drove 2500-Year Collapse."

40. Toth and Aronson, "4.2ka Event, ENSO"; Toth et al., "ENSO Drove 2500-Year Collapse."

41. Chris T. Perry and Lorenzo Alvarez-Filip, "Changing Geo-Ecological Functions of Coral Reefs in the Anthropocene," *Functional Ecology* 33, no. 6 (2019): 976–88.

42. Lauren T. Toth et al., "A 3,000-Year Lag between the Geological and Ecological Shutdown of Florida's Coral Reefs," *Global Change Biology* 24, no. 11 (2018): 5471–83.

43. Toth et al., "3,000-Year Lag."

44. Toth et al., "3,000-Year Lag"; Lauren T. Toth et al., "The Past, Present, and Future of Coral Reef Growth in the Florida Keys," *Global Change Biology* 28, no. 17 (2022): 5294–309.

45. H. Renssen et al., "Global Characterization of the Holocene Thermal Maximum," *Quaternary Science Reviews* 48 (August 10, 2012): 7–19.

46. Renssen et al.

47. Toth et al., "Past, Present, and Future"; Lauren T. Toth et al., "Climate and the Latitudinal Limits of Subtropical Reef Development," *Scientific Reports* 11, no. 1 (June 22, 2021): 13044.

48. Toth et al., "3,000-Year Lag."

49. Diego Lirman et al., "Severe 2010 Cold-Water Event Caused Unprecedented Mortality to Corals of the Florida Reef Tract and Reversed Previous Survivorship Patterns," *PLOS ONE* 6, no. 8 (August 10, 2011): e23047.

50. Stephanie Pappas, "Florida Cold Snap Devastated Coral Reefs," livescience.com, August 26, 2011, https://www.livescience.com/15799-florida-cold-devastated-coral-reefs.html.

51. Toth et al., "Past, Present, and Future."

52. Webster et al., "Response of the Great Barrier Reef."

## Chapter 7. Lying Low to Avoid Extinction

1. A good general reference about ice age climate cycles is Ruddiman, *Earth's Climate*.

2. Wood, *Reef Evolution*, 151–53.

3. Chris L. Schneider, "Marine Refugia Past, Present, and Future: Lessons from Ancient Geologic Crises for Modern Marine Ecosystem Conservation," in *Marine Conservation Paleobiology*, ed. Carrie L. Tyler and Chris L. Schneider (Cham, Switzerland: Springer International, 2018), 163–208; Gunnar Keppel et al., "Refugia: Identifying and Understanding Safe Havens for Biodiversity under Climate Change," *Global Ecology and Biogeography* 21, no. 4 (2012): 393–404; Daniel G. Gavin et al., "Climate Refugia: Joint Inference from Fossil Records, Species Distribution Models and Phylogeography," *New Phytologist* 204, no. 1 (2014): 37–54.

4. Schneider, "Marine Refugia Past, Present, and Future."

5. Schneider.

6. Keppel et al., "Refugia: Identifying and Understanding Safe Havens."

7. Schneider, "Marine Refugia Past, Present, and Future."

8. J. A. Kleypas, "Modeled Estimates of Global Reef Habitat and Carbonate Production since the Last Glacial Maximum," *Paleoceanography* 12, no. 4 (1997): 533–45.

9. William B. Ludt and Luiz A. Rocha, "Shifting Seas: The Impacts of Pleistocene Sea-Level Fluctuations on the Evolution of Tropical Marine Taxa," *Journal of Biogeography* 42, no. 1 (2015): 25–38.

10. Ludt and Rocha.

11. Ludt and Rocha.

12. Casazza, "Pleistocene Reefs of the Egyptian Red Sea."

13. Wood, *Reef Evolution*, 30.

14. Jeroen A. M. Kenter, "Carbonate Platform Flanks: Slope Angle and Sediment Fabric," *Sedimentology* 37, no. 5 (1990): 777–94; G. Michael Grammer and Robert N. Ginsburg, "Highstand versus Lowstand Deposition on Carbonate Platform Margins: Insight from Quaternary Foreslopes in the Bahamas," *Marine Geology* 103, no. 1 (January 1, 1992): 125–36.

15. Kenter, "Carbonate Platform Flanks"; Wood, *Reef Evolution*, 30; W. David Liddell and Sharon L. Ohlhorst, "Hard Substrata Community Patterns, 1–120 M, North Jamaica," *PALAIOS* 3, no. 4 (1988): 413–23; Grammer and Ginsburg, "Highstand versus Lowstand Deposition."

16. Kenter, "Carbonate Platform Flanks."

17. Kenter.

18. Ludt and Rocha, "Shifting Seas"; Kleypas, "Modeled Estimates of Global Reef Habitat."

19. Wood, *Reef Evolution*, 152.

20. Webster et al., "Response of the Great Barrier Reef."

21. Toth et al., "ENSO Drove 2500-Year Collapse."

22. Schneider, "Marine Refugia Past, Present, and Future"; Wood, *Reef Evolution*, 151–53.

23. Donald Potts and J. Jacobs, "Evolution of Reef-Building Scleractinian Corals in Turbid Environments: A Paleo-Ecological Hypothesis," in *Proceedings of the 9th International Coral Reef Symposium*, ed. M. Kasim Moosa et al. (Bali, Indonesia: International Society for Reef Studies, 2000), 249–54.

24. Madeleine J. H. van Oppen et al., "Historical and Contemporary Factors Shape the Population Genetic Structure of the Broadcast Spawning Coral, *Acropora millepora*, on the Great Barrier Reef," *Molecular Ecology* 20, no. 23 (December 2011): 4899–914.

25. van Oppen et al.

26. D. S. Portnoy et al., "Contemporary Population Structure and Post-Glacial Genetic Demography in a Migratory Marine Species, the Blacknose Shark, *Carcharhinus acronotus*," *Molecular Ecology* 23, no. 22 (2014): 5480–95.

27. Portnoy et al.

28. Wolfgang Kiessling et al., "Equatorial Decline of Reef Corals during the Last Pleistocene Interglacial," *Proceedings of the National Academy of Sciences* 109, no. 52 (December 26, 2012): 21378–83.

29. Willem Renema et al., "Are Coral Reefs Victims of Their Own Past Success?," *Science Advances* 2, no. 4 (April 22, 2016): e1500850.

30. Renema et al.; Wood, *Reef Evolution*, 152.

31. Renema et al., "Are Coral Reefs Victims?"

32. Crabbe et al., "*Acropora cervicornis*"; Crabbe et al., "*Acropora palmata*."

33. Renema et al., "Are Coral Reefs Victims?"

34. Lisa Greer et al., "Coral Gardens Reef, Belize: A Refugium in the Face of Caribbean-Wide *Acropora* spp. Coral Decline," *PLOS ONE* 15, no. 9 (September 30, 2020): e0239267.

35. Greer et al.

36. Greer et al.

37. Greer et al.

38. Greer et al.

39. I. Nagelkerken and G. van der Velde, "Relative Importance of Interlinked Mangroves and Seagrass Beds as Feeding Habitats for Juvenile Reef Fish on a Caribbean Island," *Marine Ecology*

*Progress Series* 274 (2004): 153–59; Mathias M. Igulu et al., "Mangrove Habitat Use by Juvenile Reef Fish: Meta-Analysis Reveals That Tidal Regime Matters More than Biogeographic Region," *PLOS ONE* 9, no. 12 (December 31, 2014): e114715.

40. Katherine C. Ewel, Robert R. Twilley, and Jin Eong Ong, "Different Kinds of Mangrove Forests Provide Different Goods and Services," *Global Ecology and Biogeography Letters* 7, no. 1 (1998): 83–94.

41. Ewel, Twilley, and Ong.

42. Mark Easter, *The Blue Plate: A Food Lover's Guide to Climate Chaos* (Ventura, CA: Patagonia, 2024).

43. "A Record Amount of Seaweed Is Choking Shores in the Caribbean," National Public Radio, Environment, August 3, 2022, https://www.npr.org/2022/08/03/1115383385/seaweed-record-caribbean.

44. "Understanding the Spread of Sargassum in the Caribbean," UN Environmental Programme, 2015, https://www.unep.org/cep/resources/factsheet/understanding-spread-sargassum-caribbean.

45. Brigitta I. van Tussenbroek et al., "Severe Impacts of Brown Tides Caused by *Sargassum* spp. on Near-Shore Caribbean Seagrass Communities," *Marine Pollution Bulletin* 122, no. 1 (September 15, 2017): 272–81.

46. Esteban A. Agudo-Adriani et al., "Colony Geometry and Structural Complexity of the Endangered Species *Acropora cervicornis* Partly Explains the Structure of Their Associated Fish Assemblage," *PeerJ* 4 (April 4, 2016): e1861.

47. Gardner et al., "Long-Term Region-Wide Declines"; Jackson et al., *Status and Trends*.

48. Lisa Greer et al., "Coral Gardens, Belize: Quantifying Acroporid Coral Survivors," GSA Connects, 2022, https://gsa.confex.com/gsa/2022AM/meetingapp.cgi/Paper/379550.

49. M. McField et al., "2022 Mesoamerican Reef Report Card: An Evaluation of Ecosystem Health," Healthy Reefs Initiative, accessed September 16, 2024, https://healthyreefs.org/report-cards/.

50. Lisa Greer et al., "Coral Gardens Reef, Belize: An *Acropora* spp. Refugium under Threat in a Warming World," *PLOS ONE* 18, no. 2 (February 8, 2023): e0280852.

51. Greer et al., "Coral Gardens Reef, Belize: A Refugium in the Face of Caribbean-Wide *Acropora* spp. Coral Decline."

52. Greer et al.

53. Greer et al.

54. Greer et al.

55. Greer et al., "Coral Gardens, Belize: Quantifying Acroporid Coral Survivors."

56. Ove Hoegh-Guldberg et al., "Coral Reefs in Peril in a Record-Breaking Year," *Science* 382, no. 6676 (December 15, 2023): 1238–40.

57. Elizabeth Claire Alberts, "An El Niño Is Forecast for 2023. How Much Coral Will Bleach This Time?," Mongabay Environmental News, February 2, 2023, https://news.mongabay.com/2023/02/an-el-nino-is-forecast-for-2023-how-much-coral-will-bleach-this-time/.

58. "Belize Regional Products," NOAA Coral Reef Watch, accessed December 30, 2023, https://coralreefwatch.noaa.gov/product/vs/gauges/belize.php.

59. Knowlton et al., "Rebuilding Coral Reefs."

60. *Special Report: Global Warming of 1.5°C*, Intergovernmental Panel on Climate Change, 2018, https://www.ipcc.ch/sr15/; Knowlton et al., "Rebuilding Coral Reefs."

61. Javid Kavousi and Gunnar Keppel, "Clarifying the Concept of Climate Change Refugia for Coral Reefs," *ICES Journal of Marine Science* 75, no. 1 (January 1, 2018): 43–49.

62. Schneider, "Marine Refugia Past, Present, and Future."

63. Hawthorne L. Beyer et al., "Risk-Sensitive Planning for Conserving Coral Reefs under Rapid Climate Change," *Conservation Letters* 11, no. 6 (2018): e12587.

64. "50 Reefs: A Global Plan to Save Coral Reefs," The Ocean Agency, accessed June 8, 2023, https://www.50reefs.org.

65. Adele M. Dixon et al., "Future Loss of Local-Scale Thermal Refugia in Coral Reef Ecosystems," *PLOS Climate* 1, no. 2 (February 1, 2022): e0000004.

66. "NOAA Coral Reef Watch Tutorial," accessed June 9, 2023, https://coralreefwatch.noaa.gov/product/5km/tutorial/crw10a_dhw_product.php.

67. McWhorter et al., "Climate Refugia on the Great Barrier Reef."

68. Iliana Chollett and Peter J. Mumby, "Reefs of Last Resort: Locating and Assessing Thermal Refugia in the Wider Caribbean," *Biological Conservation* 167 (November 1, 2013): 179–86.

69. B. Riegl and W. E. Piller, "Possible Refugia for Reefs in Times of Environmental Stress," *International Journal of Earth Sciences* 92, no. 4 (September 1, 2003): 520–31; Bernhard Riegl and Werner Piller, "Upwelling Areas as Possible Refugia for Reefs in Times of Rising SST? Further Evidence from the Caribbean and Indian Oceans," in *Proceedings of the 9th International Coral Reef Symposium*, ed. M. Kasim Moosa et al. (Bali, Indonesia: International Society for Reef Studies, 2000), 315–20.

70. McWhorter et al., "Climate Refugia on the Great Barrier Reef."

71. McWhorter et al.

72. McWhorter et al.

73. McWhorter et al.

74. McWhorter et al.

75. M. Wall et al., "Large-Amplitude Internal Waves Benefit Corals during Thermal Stress," *Proceedings of the Royal Society B: Biological Sciences* 282, no. 1799 (January 22, 2015): 20140650.

76. Wall et al.

77. Wall et al.

78. Scott D. Bachman et al., "A Global Atlas of Potential Thermal Refugia for Coral Reefs Generated by Internal Gravity Waves," *Frontiers in Marine Science* 9 (2022).

79. Bachman et al.

80. Bachman et al.

81. P. Bongaerts et al., "Assessing the 'Deep Reef Refugia' Hypothesis: Focus on Caribbean Reefs," *Coral Reefs* 29, no. 2 (June 1, 2010): 309–27.

82. Riegl and Piller, "Possible Refugia for Reefs."

83. Bongaerts et al., "Assessing the 'Deep Reef Refugia' Hypothesis."

84. *Mesophotic Coral Ecosystems: A Lifeboat for Coral Reefs?*, United Nations Environment Programme and GRID-Arendal, May 2016, http://www.unep.org/resources/report/mesophotic-coral-ecosystems-lifeboat-coral-reefs.

85. *Mesophotic Coral Ecosystems*, 10.

86. Tyler B. Smith et al., "Caribbean Mesophotic Coral Ecosystems Are Unlikely Climate Change Refugia," *Global Change Biology* 22, no. 8 (2016): 2756–65; Juliano Morais and Bráulio A. Santos, "Limited Potential of Deep Reefs to Serve as Refuges for Tropical Southwestern Atlantic Corals," *Ecosphere* 9, no. 7 (2018): e02281.

87. *Mesophotic Coral Ecosystems*, 71.

88. McWhorter et al., "Climate Refugia on the Great Barrier Reef."

89. Shannon Sully and Robert van Woesik, "Turbid Reefs Moderate Coral Bleaching under Climate-Related Temperature Stress," *Global Change Biology* 26, no. 3 (March 2020): 1367–73; Miguel Mies et al., "South Atlantic Coral Reefs Are Major Global Warming Refugia and Less Susceptible to Bleaching," *Frontiers in Marine Science* 7 (2020).

90. Chris Cacciapaglia and Robert van Woesik, "Climate-Change Refugia: Shading Reef Corals by Turbidity," *Global Change Biology* 22, no. 3 (March 1, 2016): 1145–54; Sully and van Woesik, "Turbid Reefs Moderate Coral Bleaching."

91. Cacciapaglia and van Woesik, "Climate-Change Refugia."

92. McWhorter et al., "Climate Refugia on the Great Barrier Reef."

## Chapter 8. Designing Reefs That Can Survive Us

1. Carl V. Miller et al., "Exploratory Treatments for Stony Coral Tissue Loss Disease: Pillar Coral (*Dendrogyra cylindrus*)," NOAA Technical Memorandum NOS NCCOS 245 and CRCP 37, 2020, 78.

2. Adele Irwin et al., "Age and Intraspecific Diversity of Resilient *Acropora* Communities in Belize," *Coral Reefs* 36, no. 4 (December 1, 2017): 1111–20.

3. Brooks Hays, "Coral Survey Reveals 5,000-Year-Old Genotypes," United Press International, 2016, https://www.upi.com/Science_News/2016/11/30/Coral-survey-reveals-5000-year-old-genotypes/9181480516513/.

4. Lisa Boström-Einarsson et al., "Coral Restoration—A Systematic Review of Current Methods, Successes, Failures and Future Directions," *PLOS ONE* 15, no. 1 (January 30, 2020): e0226631.

5. "Fragments of Hope, Belize: About Us," accessed June 8, 2023, https://fragmentsofhope.org/home-2/about-us/.

6. Aronson et al., "Expanding Scale of Species Turnover."

7. Muraoka et al., "Historical Declines in Parrotfish."

8. "Laughing Bird Caye National Park Management Plan 2011–2016: A Component of Belize's World Heritage Site," NOAA, accessed June 8, 2023, https://repository.library.noaa.gov/view/noaa/820.

9. "Belize Barrier Reef Reserve System," UNESCO World Heritage Centre, accessed July 4, 2023, https://whc.unesco.org/en/list/764/.

10. Boström-Einarsson et al., "Coral Restoration."

11. Christopher A. Page, Erinn M. Muller, and David E. Vaughan, "Microfragmenting for the Successful Restoration of Slow Growing Massive Corals," *Ecological Engineering* 123 (November 1, 2018): 86–94.

12. Toth et al., "Past, Present, and Future."

13. Toth et al.

14. "Restoring Seven Iconic Reefs: A Mission to Recover the Coral Reefs of the Florida Keys," NOAA Fisheries, March 8, 2022, https://www.fisheries.noaa.gov/southeast/habitat-conservation/restoring-seven-iconic-reefs-mission-recover-coral-reefs-florida-keys; "NOAA Launches Mission: Iconic Reefs to Save Florida Keys Coral Reefs," NOAA National Marine Sanctuaries, accessed June 8, 2023, https://sanctuaries.noaa.gov/news/dec19/noaa-launches-mission-iconic-reefs-to-save-florida-keys-coral-reefs.html.

15. Aaron R. Pilnick et al., "A Novel System for Intensive *Diadema antillarum* Propagation as a Step towards Population Enhancement," *Scientific Reports* 11, no. 1 (May 27, 2021): 11244.

16. Toth et al., "Past, Present, and Future."

17. "Coping with the 2023 Coral Bleaching: Prepare, Manage, Monitor, and Recover," Coral Restoration Consortium, 2023, https://www.youtube.com/watch?v=yiX2XOxUFak.

18. Michael Childress, "The Heroic Effort to Save Florida's Coral Reef from Extreme Ocean Heat as Corals Bleach across the Caribbean," The Conversation, August 9, 2023, http://theconversation.com/the-heroic-effort-to-save-floridas-coral-reef-from-extreme-ocean-heat-as-corals-bleach-across-the-caribbean-210974.

19. "Raising Coral Costa Rica: A Human-Coral Symbiosis," Raising Coral Costa Rica, accessed June 8, 2023, https://www.raisingcoral.org.

20. Boström-Einarsson et al., "Coral Restoration."

21. Ken Anthony et al., "New Interventions Are Needed to Save Coral Reefs," *Nature Ecology & Evolution* 1, no. 10 (October 2017): 1420–22.

22. Boström-Einarsson et al., "Coral Restoration."

23. Beth Alexander, "Artificial Reefs: Environmental Solution or Problem," *Scuba Diver Life* (blog), July 30, 2014, https://scubadiverlife.com/artificial-reefs-environmental-solution-problem/.

24. Avery B. Paxton et al., "Meta-Analysis Reveals Artificial Reefs Can Be Effective Tools for Fish Community Enhancement but Are Not One-Size-Fits-All," *Frontiers in Marine Science* 7 (2020); Thiony Simon, Jean-Christophe Joyeux, and Hudson T. Pinheiro, "Fish Assemblages on Shipwrecks and Natural Rocky Reefs Strongly Differ in Trophic Structure," *Marine Environmental Research* 90 (September 1, 2013): 55–65.

25. "National Guidance: Best Management Practices for Preparing Vessels Intended to Create Artificial Reefs," US Environmental Protection Agency, 2006, https://www.epa.gov/sites/default/files/2015-09/documents/artificialreefguidance.pdf.

26. Juliet Eilperin, "Artificial Reef Projects Raise Environmental Questions," *Washington Post*, July 17, 2011, https://www.washingtonpost.com/national/health-science/artificial-reef-projects-raise-environmental-questions/2011/07/13/gIQAmIRRKI_story.html; Jun-Ki Choi et al., "Economic and Environmental Perspectives of End-of-Life Ship Management," *Resources, Conservation and Recycling* 107 (February 1, 2016): 82–91.

27. "Artificial Reef Made out of Tires Causing Big Mess off Fort Lauderdale Coast," 2019, https://www.youtube.com/watch?v=YaD-ewI81R4.

28. Claudia E. L. Hill, Myrsini M. Lymperaki, and Bert W. Hoeksema, "A Centuries-Old Manmade Reef in the Caribbean Does Not Substitute Natural Reefs in Terms of Species Assemblages and Interspecific Competition," *Marine Pollution Bulletin* 169 (August 1, 2021): 112576.

29. Maxwell B. Kaplan and T. Aran Mooney, "Coral Reef Soundscapes May Not Be Detectable Far from the Reef," *Scientific Reports* 6, no. 1 (August 23, 2016): 31862; M. B. Kaplan et al., "Coral Reef Species Assemblages Are Associated with Ambient Soundscapes," *Marine Ecology Progress Series* 533 (August 6, 2015).

30. Kaplan et al., "Coral Reef Species Assemblages."

31. Kaplan and Mooney, "Coral Reef Soundscapes"; Maxwell B. Kaplan et al., "Acoustic and Biological Trends on Coral Reefs off Maui, Hawaii," *Coral Reefs* 37, no. 1 (March 1, 2018): 121–33.

32. "Can We Use Sound to Build Back Reefs?," Woods Hole Oceanographic Institution, 2022, https://www.youtube.com/watch?v=VIaGVetGnRg.

33. "Can We Use Sound to Build Back Reefs?"

34. "MARS: Modular Artificial Reef Structure," Reef Design Lab, accessed June 8, 2023, https://www.reefdesignlab.com/mars.

35. "MARS: Modular Artificial Reef Structure."

36. "MARS (Modular Artificial Reef Structure) Designed by Alex Goad," 2018, https://www.youtube.com/watch?v=BMCfiLnncg8; "MARS 2022 Summer Island Maldives_2," 2022, https://www.youtube.com/watch?v=Z0jvjoFWwUY; "Summer Island's 3D Printed Artificial Coral Reef," 2018, https://www.youtube.com/watch?v=Nc5SBCw_DJo.

37. *Broken Nature*, Museum of Modern Art, accessed June 8, 2023, https://www.moma.org/calendar/exhibitions/5220.

38. Susan L. Williams et al., "Large-Scale Coral Reef Rehabilitation after Blast Fishing in Indonesia," *Restoration Ecology* 27, no. 2 (2019): 447–56.

39. Williams et al.

40. "Reef Revitalization," University of Miami, College of Arts and Sciences, accessed June 8, 2023, https://news.miami.edu/as/stories/2022/06/reef-revitalization.html.

41. "Reef Revitalization."

42. Robert C. Jones Jr. and Janette Neuwahl Tannen, "Unique Hybrid Reefs Deployed off Miami Beach," University of Miami, accessed September 4, 2024, https://news.miami.edu/stories/2023/03/unique-hybrid-reefs-deployed-off-miami-beach.html.

43. "Great Barrier Reef Coral Bleaching to Be Tackled in Cloud-Brightening Experiment," ABC News (Australia), 2020, https://www.youtube.com/watch?v=o8JkN7XbdZk.

44. Graham Readfearn, "Coalition Backs 'Cloud-Brightening' Trial on Great Barrier Reef to Tackle Global Heating," *Guardian*, July 14, 2020, https://www.theguardian.com/environment/2020/jul/15/coalition-backs-cloud-brightening-trial-on-great-barrier-reef-to-tackle-global-heating.

45. "How Cloud Brightening Protects Australia's Great Barrier Reef," 2021, https://www.youtube.com/watch?v=qjya_arTWjs.

46. *Climate Change 2013: The Physical Science Basis*, Intergovernmental Panel on Climate Change, 2013, https://www.ipcc.ch/report/ar5/wg1/.

47. Jeff Tollefson, "Can Artificially Altered Clouds Save the Great Barrier Reef?," *Nature* 596, no. 7873 (August 25, 2021): 476–78.

48. Tollefson.

49. Joan Kleypas et al., "Designing a Blueprint for Coral Reef Survival," *Biological Conservation* 257 (May 1, 2021): 109107.

50. James E. Palardy, Lisa J. Rodrigues, and Andréa G. Grottoli, "The Importance of Zooplankton to the Daily Metabolic Carbon Requirements of Healthy and Bleached Corals at Two Depths," *Journal of Experimental Marine Biology and Ecology* 367, no. 2 (December 15, 2008): 180–88.

51. Andréa Grottoli, "The Future of Coral Reefs," Rachel Carson Lecture, AGU Fall Meeting, December 15, 2021.

## Chapter 9. Survival of the Heat Tolerant

1. Peter W. Glynn, Derek P. Manzello, and Ian C. Enochs, eds., *Coral Reefs of the Eastern Tropical Pacific: Persistence and Loss in a Dynamic Environment*, vol. 8 of *Coral Reefs of the World* (Dordrecht: Springer Netherlands, 2017).

2. Glynn, Manzello, and Enochs.

3. Charles Darwin, *On the Origin of Species: A Facsimile of the First Edition* (Cambridge, MA: Harvard University Press, 2001), chap. 1; "Darwin Correspondence Project," accessed June 9, 2023, https://www.darwinproject.ac.uk/; James A. Secord, "Nature's Fancy: Charles Darwin and the Breeding of Pigeons," *Isis* 72, no. 2 (1981): 163–86.

4. Secord, "Nature's Fancy."

5. "Survival of the Fittest," Darwin Correspondence Project, University of Cambridge, accessed September 16, 2024, https://www.darwinproject.ac.uk/commentary/survival-fittest.

6. Darwin, *On the Origin of Species*.

7. Janet Browne, "Wallace and Darwin," *Current Biology* 23, no. 24 (December 16, 2013): R1071–72.

8. Browne.

9. Browne.

10. Charles Darwin et al., "On the Tendency of Species to Form Varieties; and on the Perpetuation of Varieties and Species by Natural Means of Selection," *Zoological Journal of the Linnean Society* 3 (1858): 46–50.

11. Browne, "Wallace and Darwin."

12. Nikolaus B. M. Császár et al., "Estimating the Potential for Adaptation of Corals to Climate Warming," *PLOS ONE* 5, no. 3 (March 18, 2010): e9751.

13. Michael L. Berumen et al., "Corals of the Red Sea," in *Coral Reefs of the Red Sea*, ed. Christian R. Voolstra and Michael L. Berumen, vol. 11 of *Coral Reefs of the World* (Berlin: Springer International Publishing, 2019), 123–55.

14. Andréa G. Grottoli, Dan Tchernov, and Gidon Winters, "Physiological and Biogeochemical Responses of Super-Corals to Thermal Stress from the Northern Gulf of Aqaba, Red Sea," *Frontiers in Marine Science* 4 (2017).

15. Maoz Fine, Hezi Gildor, and Amatzia Genin, "A Coral Reef Refuge in the Red Sea," *Global Change Biology* 19, no. 12 (2013): 3640–47.

16. Fine, Gildor, and Genin.

17. Fine, Gildor, and Genin.

18. Fine, Gildor, and Genin.

19. Grottoli, Tchernov, and Winters, "Physiological and Biogeochemical Responses."

20. "Satellites and Bleaching," NOAA Coral Reef Watch Tutorial, accessed June 9, 2023, https://coralreefwatch.noaa.gov/product/5km/tutorial/crw10a_dhw_product.php.

21. Grottoli, Tchernov, and Winters, "Physiological and Biogeochemical Responses."

22. Grottoli, Tchernov, and Winters.

23. Jonah Mandel, "In the Red Sea, Coral Reefs Can Take the Heat of Climate Change," accessed June 9, 2023, https://phys.org/news/2017-06-red-sea-coral-reefs-climate.html.

24. Fine, Gildor, and Genin, "Coral Reef Refuge in the Red Sea."

25. "Coral Reefs: From Climate Victims to Survivors," 2022, https://www.youtube.com/watch?v=lN6WPrLK6Gs.

26. Madeleine J. H. van Oppen et al., "Building Coral Reef Resilience through Assisted Evolution," *Proceedings of the National Academy of Sciences* 112, no. 8 (February 24, 2015): 2307–13.

27. Daniel J. Barshis et al., "Genomic Basis for Coral Resilience to Climate Change," *Proceedings of the National Academy of Sciences* 110, no. 4 (January 22, 2013): 1387–92.

28. van Oppen et al., "Building Coral Reef Resilience"; Lukas B. DeFilippo et al., "Assessing the Potential for Demographic Restoration and Assisted Evolution to Build Climate Resilience in Coral Reefs," *Ecological Applications* 32, no. 7 (2022): e2650.

29. Emily J. Howells et al., "Enhancing the Heat Tolerance of Reef-Building Corals to Future Warming," *Science Advances* 7, no. 34 (August 20, 2021): eabg6070.

30. DeFilippo et al., "Assessing the Potential."

31. "Responding to the 2023 Caribbean Coral Bleaching Event—September 11, 2023 CCT Webinar," 2023, https://www.youtube.com/watch?v=7KlCuBXSrHw.

32. "Responding to the 2023 Caribbean Coral Bleaching Event."

33. "Responding to the 2023 Caribbean Coral Bleaching Event."

34. Iliana B. Baums et al., "Considerations for Maximizing the Adaptive Potential of Restored Coral Populations in the Western Atlantic," *Ecological Applications* 29, no. 8 (2019): e01978.

35. Baums et al.

36. Kent E. Carpenter et al., "One-Third of Reef-Building Corals Face Elevated Extinction Risk from Climate Change and Local Impacts," *Science* 321, no. 5888 (July 25, 2008): 560–63.

37. Barshis et al., "Genomic Basis for Coral Resilience."

38. Monica Edwards, Laurie Abadie, and Kelli Mars, "NASA Twins Study Confirms Preliminary Findings," NASA, January 31, 2018, https://www.nasa.gov/humans-in-space/nasa-twins-study-confirms-preliminary-findings/.

39. Romain Savary et al., "Fast and Pervasive Transcriptomic Resilience and Acclimation of Extremely Heat-Tolerant Coral Holobionts from the Northern Red Sea," *Proceedings of the National Academy of Sciences* 118, no. 19 (May 11, 2021): e2023298118; Ana M. Palacio-Castro et al., "Increased Dominance of Heat-Tolerant Symbionts Creates Resilient Coral Reefs in Near-Term Ocean Warming," *Proceedings of the National Academy of Sciences* 120, no. 8 (February 21, 2023): e2202388120; Victoria Barker, "Exceptional Thermal Tolerance of Coral Reefs in American Samoa: A Review," *Current Climate Change Reports* 4, no. 4 (December 1, 2018): 417–27.

40. Hollie M. Putnam and Ruth D. Gates, "Preconditioning in the Reef-Building Coral *Pocillopora damicornis* and the Potential for Trans-Generational Acclimatization in Coral Larvae

under Future Climate Change Conditions," *Journal of Experimental Biology* 218, no. 15 (August 2015): 2365–72.

41. Hollie M. Putnam et al., "The Vulnerability and Resilience of Reef-Building Corals," *Current Biology* 27, no. 11 (June 5, 2017): R528–40.

42. Raquel S. Peixoto et al., "Beneficial Microorganisms for Corals (BMC): Proposed Mechanisms for Coral Health and Resilience," *Frontiers in Microbiology* 8 (2017).

43. Luke K. Ursell et al., "Defining the Human Microbiome," *Nutrition Reviews* 70, suppl. 1 (August 2012): S38–44.

44. Yong, *I Contain Multitudes*, 105.

45. Linda L. Blackall, Bryan Wilson, and Madeleine J. H. van Oppen, "Coral—the World's Most Diverse Symbiotic Ecosystem," *Molecular Ecology* 24, no. 21 (2015): 5330–47.

46. Peixoto et al., "Beneficial Microorganisms for Corals"; Heru Kusdianto et al., "Microbiomes of Healthy and Bleached Corals during a 2016 Thermal Bleaching Event in the Upper Gulf of Thailand," *Frontiers in Marine Science* 8 (2021).

47. Peixoto et al., "Beneficial Microorganisms for Corals."

48. Rosales et al., "Microbiome Differences."

49. Kusdianto et al., "Microbiomes of Healthy and Bleached Corals"; Andréa G. Grottoli et al., "Coral Physiology and Microbiome Dynamics under Combined Warming and Ocean Acidification," *PLOS ONE* 13, no. 1 (January 16, 2018): e0191156.

50. Peixoto et al., "Beneficial Microorganisms for Corals."

51. Hannah E. Epstein, Gergely Torda, and Madeleine J. H. van Oppen, "Relative Stability of the *Pocillopora acuta* Microbiome throughout a Thermal Stress Event," *Coral Reefs* 38, no. 2 (April 1, 2019): 373–86; Grottoli et al., "Coral Physiology and Microbiome Dynamics."

52. Grottoli et al., "Coral Physiology and Microbiome Dynamics."

53. Peixoto et al., "Beneficial Microorganisms for Corals"; T. Doering, M. J. van Oppen, and L. L. Blackall, "Advancing Coral Microbiome Manipulation to Build Long-Term Climate Resilience," *Microbiology Australia* 44, no. 1 (2023): 36–40.

54. Doering, van Oppen, and Blackall, "Advancing Coral Microbiome Manipulation."

55. Jos C. Mieog et al., "The Roles and Interactions of Symbiont, Host and Environment in Defining Coral Fitness," *PLOS ONE* 4, no. 7 (July 24, 2009): e6364.

56. Rachel N. Silverstein, Ross Cunning, and Andrew C. Baker, "Change in Algal Symbiont Communities after Bleaching, Not Prior Heat Exposure, Increases Heat Tolerance of Reef Corals," *Global Change Biology* 21, no. 1 (January 2015): 236–49.

57. R. Cunning et al., "Growth Tradeoffs Associated with Thermotolerant Symbionts in the Coral *Pocillopora damicornis* Are Lost in Warmer Oceans," *Coral Reefs* 34, no. 1 (March 1, 2015): 155–60; Ross Cunning, "Will Coral Reefs Survive by Adaptive Bleaching?," *Emerging Topics in Life Sciences* 6, no. 1 (March 14, 2022): 11–15.

58. Palacio-Castro et al., "Increased Dominance of Heat-Tolerant Symbionts."

59. Palacio-Castro et al.

60. T. A. Oliver and S. R. Palumbi, "Do Fluctuating Temperature Environments Elevate Coral Thermal Tolerance?," *Coral Reefs* 30, no. 2 (June 1, 2011): 429–40; Stephen R. Palumbi et al., "Mechanisms of Reef Coral Resistance to Future Climate Change," *Science* 344, no. 6186 (May 23, 2014): 895–98; Barker, "Exceptional Thermal Tolerance."

61. Barker, "Exceptional Thermal Tolerance."

62. Palumbi et al., "Mechanisms of Reef Coral Resistance"; Oliver and Palumbi, "Do Fluctuating Temperature Environments?"; Barker, "Exceptional Thermal Tolerance."

63. Palumbi et al., "Mechanisms of Reef Coral Resistance"; Oliver and Palumbi, "Do Fluctuating Temperature Environments?"

64. Palumbi et al., "Mechanisms of Reef Coral Resistance"; Oliver and Palumbi, "Do Fluctuating Temperature Environments?"

65. Palumbi et al., "Mechanisms of Reef Coral Resistance"; Oliver and Palumbi, "Do Fluctuating Temperature Environments?"

66. Oliver and Palumbi, "Do Fluctuating Temperature Environments?"

67. Oliver and Palumbi.

68. Palumbi et al., "Mechanisms of Reef Coral Resistance."

69. Palumbi et al.

70. Palumbi et al.

71. Luke Thomas et al., "Mechanisms of Thermal Tolerance in Reef-Building Corals across a Fine-Grained Environmental Mosaic: Lessons from Ofu, American Samoa," *Frontiers in Marine Science* 4 (2018), https://www.frontiersin.org/articles/10.3389/fmars.2017.00434.

72. Barshis et al., "Genomic Basis for Coral Resilience."

73. Palumbi et al., "Mechanisms of Reef Coral Resistance."

74. Thomas et al., "Mechanisms of Thermal Tolerance."

75. Stephen R. Palumbi and Anthony R. Palumbi, *The Extreme Life of the Sea* (Princeton, NJ: Princeton University Press, 2014), 118.

76. Aryan Safaie et al., "High Frequency Temperature Variability Reduces the Risk of Coral Bleaching," *Nature Communications* 9, no. 1 (April 26, 2018): 1671; Thomas et al., "Mechanisms of Thermal Tolerance."

77. D. M. Thompson and R. van Woesik, "Corals Escape Bleaching in Regions That Recently and Historically Experienced Frequent Thermal Stress," *Proceedings of the Royal Society B: Biological Sciences* 276, no. 1669 (May 27, 2009): 2893–901.

78. Palacio-Castro et al., "Increased Dominance of Heat-Tolerant Symbionts."

### Chapter 10. The Anthropocene Coral Paradox and the Future of Reefs

1. Greer et al., "Coral Gardens Reef, Belize: A Refugium in the Face of Caribbean-Wide *Acropora* spp. Coral Decline."

2. "Human Activity Devastating Marine Species from Mammals to Corals - IUCN Red List," IUCN Press Release, December 9, 2022, https://www.iucn.org/press-release/202212/human-activity-devastating-marine-species-mammals-corals-iucn-red-list; F. Cavada-Blanco et al., "IUCN Red List of Threatened Species: *Dendrogyra cylindrus*," IUCN Red List of Threatened Species, 2022, https://www.iucnredlist.org/en.

3. Karen L. Neely et al., "Rapid Population Decline of the Pillar Coral *Dendrogyra cylindrus* along the Florida Reef Tract," *Frontiers in Marine Science* 8 (2021); "Pillar Coral Was Already Rare on Florida Reefs. Now Biologists Say It's 'Extinct,'" WLRN, May 19, 2021, https://www

.wlrn.org/news/2021-05-19/pillar-coral-was-already-rare-on-florida-reefs-now-biologists-say-its-extinct.

4. Cassie Freund, "Pillar Corals Now Considered Critically Endangered," Frost Science, 2023, https://www.frostscience.org/pillar-corals-now-considered-critically-endangered/.

5. Freund.

6. "Human Activity Devastating Marine Species."

7. "Human Activity Devastating Marine Species."

8. Andrew Frederick Johnson, "Solving 'Darwin's Paradox': Why Coral Island Hotspots Exist in an Oceanic Desert," The Conversation, February 18, 2016, http://theconversation.com/solving-darwins-paradox-why-coral-island-hotspots-exist-in-an-oceanic-desert-54719.

9. Pandolfi and Jackson, "Community Structure of Pleistocene Coral Reefs"; Pandolfi and Jackson, "Ecological Persistence Interrupted."

10. Pandolfi and Jackson, "Ecological Persistence Interrupted"; Jackson, "Pleistocene Perspectives"; Pandolfi, "Limited Membership."

11. Webster et al., "Response of the Great Barrier Reef."

12. Greer et al., "Coral Gardens Reef, Belize: A Refugium in the Face of Caribbean-Wide *Acropora* spp. Coral Decline."

13. Carpenter et al., "One-Third of Reef-Building Corals."

14. "What Is the Sixth Mass Extinction and What Can We Do about It?," World Wildlife Fund, accessed January 19, 2024, https://www.worldwildlife.org/stories/what-is-the-sixth-mass-extinction-and-what-can-we-do-about-it.

15. Hoegh-Guldberg et al., "Coral Reefs in Peril."

16. C. Mark Eakin, Hugh P. A. Sweatman, and Russel E. Brainard, "The 2014–2017 Global-Scale Coral Bleaching Event: Insights and Impacts," *Coral Reefs* 38, no. 4 (August 1, 2019): 539–45.

17. "Global Coral Bleaching Event Likely Ending," NOAA, June 19, 2017, https://www.noaa.gov/media-release/global-coral-bleaching-event-likely-ending.

18. James Davis Reimer et al., "The Fourth Global Coral Bleaching Event: Where Do We Go from Here?," *Coral Reefs* 43, no. 4 (August 1, 2024): 1121–25.

19. Eakin, Sweatman, and Brainard, "2014–2017 Global-Scale Coral Bleaching Event."

20. Eakin, Sweatman, and Brainard.

21. Palacio-Castro et al., "Increased Dominance of Heat-Tolerant Symbionts."

22. Eakin, Sweatman, and Brainard, "2014–2017 Global-Scale Coral Bleaching Event."

23. Eakin, Sweatman, and Brainard.

24. William F. Precht et al., "Unprecedented Disease-Related Coral Mortality in Southeastern Florida," *Scientific Reports* 6, no. 1 (August 10, 2016): 31374.

25. Knowlton et al., "Rebuilding Coral Reefs."

26. Hoegh-Guldberg et al., "Coral Reefs in Peril."

27. Hannah Ritchie, Max Roser, and Pablo Rosado, "$CO_2$ and Greenhouse Gas Emissions," Our World in Data, May 11, 2020, https://ourworldindata.org/emissions-by-sector.

28. "Carbonate Evolution," *Oil & Gas Middle East*, August 13, 2008, https://www.oilandgasmiddleeast.com/products-services/article-4852-carbonate-evolution.

29. "The Paris Agreement," United Nations Climate Action, accessed June 9, 2023, https://www.un.org/en/climatechange/paris-agreement.

30. *Special Report: Global Warming of 1.5°C*, 229–30.
31. *Special Report: Global Warming of 1.5°C*, 229–30.
32. Cornwall et al., "Global Declines in Coral Reef."
33. Cornwall et al.
34. Kleypas et al., "Designing a Blueprint."
35. Kleypas et al.
36. "Nations Must Go Further Than Current Paris Pledges or Face Global Warming of 2.5–2.9°C," UN Environment Programme, November 20, 2023, https://www.unep.org/news-and-stories/press-release/nations-must-go-further-current-paris-pledges-or-face-global-warming.
37. "Nations Must Go Further."
38. "Inadequate Progress on Climate Action Makes Rapid Transformation of Societies Only Option - UNEP," UN Environment Programme, October 27, 2022, http://www.unep.org/news-and-stories/press-release/inadequate-progress-climate-action-makes-rapid-transformation.
39. Knowlton et al., "Rebuilding Coral Reefs."
40. Steneck et al., "Managing Recovery Resilience."
41. Knowlton, "Ocean Optimism."
42. Knowlton.
43. Knowlton.
44. Abel Valdivia, Shaye Wolf, and Kieran Suckling, "Marine Mammals and Sea Turtles Listed under the U.S. Endangered Species Act Are Recovering," *PLOS ONE* 14, no. 1 (January 16, 2019): e0210164.
45. Knowlton et al., "Rebuilding Coral Reefs."
46. Knowlton et al.
47. Knowlton et al.
48. Juli Berwald, "The Mystery of the Healthy Coral Reef," *Nautilus*, January 19, 2023, https://nautil.us/the-mystery-of-the-healthy-coral-reef-258494/.
49. Enric Sala, "Once Devastated, These Pacific Reefs Have Seen an Amazing Rebirth," *National Geographic*, October 11, 2022, https://www.nationalgeographic.com/magazine/article/once-devastated-these-pacific-reefs-have-seen-an-amazing-rebirth-feature.
50. A. Hooijer and R. Vernimmen, "Global LiDAR Land Elevation Data Reveal Greatest Sea-Level Rise Vulnerability in the Tropics," *Nature Communications* 12, no. 1 (June 29, 2021): 3592.
51. Hooijer and Vernimmen.
52. Freund, "Pillar Corals."

# INDEX

acclimatization, 204, 209–10, 212, 224, 226
*Acropora cervicornis. See* staghorn coral
*Acropora* corals, 87–88, 148–49, 166–67, 171, 176
*Acropora hyacinthus. See* tabletop coral
*Acropora millepora*, 147
*Acropora palmata. See* elkhorn coral
*Acropora prolifera*, 148–49. *See also* prolifera coral
adaptation: of cauliflower coral, 207; evidence of, 207; experiment regarding, 200; future of, 226; human-assisted evolution for, 201; map of, 199; of peppered moths, 196–97; prediction regarding, 226; at the Red Sea, 197–98; timeframe for, 197, 224; for water temperature, 195. *See also* natural selection
*Aktuo-paläontologie* (actualistic paleontology or actuopaleontology), 83
algae, 21, 22. *See also specific types*
American Museum of Natural History, 101
Andersen, Inger, 224
Andros Island, 21
Anthropocene, 7, 222
Aoki, Nadège, 187
aquaculture methodology, 182
Aronson, Rich, 89, 95, 174
artificial reef, 185–86, 188–89
atolls: Charles Darwin and, 26, 43, 44, 45, 47; defined, 51; description of, 42; Enewetak and Bikini Atolls, 52; forming of, 41, 42, 45, 46, 47; hazards of, 42; limestone formation and, 44–45, 143; overview of, 26; size of, 42; theories regarding, 50–51; volcanoes and, 45, 51
Australia, 130, 132. *See also* Great Barrier Reef (Australia)
Australian Institute of Marine Science, 99

Bab el-Mandeb, sea level changes at, 198
Bachman, Scott, 164
backreef zone, 59, 60, 61, 151, 174, 208
bacteria, 19, 65, 95, 197, 205
Bahamas/Bahamian reefs, 21, 25, 28–29, 121, 125–26, 143. *See also specific locations*
banana production, 107
Barbados, 27, 66, 132, 133–34, 218
barrier reefs, 25–26, 76
baseline shift, 175–76, 230
Bay of Concepción, 43
beachrock, 35
Belize, 115, 157–58, 171, 174. *See also specific locations*
Belize Barrier Reef Reserve System, 179
Berwald, Juli, 230
Bikini Atolls, 52. *See also* atolls
bioerosion, 86, 220. *See also* erosion
Biscayne National Park, 76
blacknose sharks, in refugium, 147–48. *See also* sharks
Blanchon, Paul, 144
bluehead wrasse, 3
Bocas del Toro, Panama, 107
Bood, Nadia, 178–79

267

borings: at Cockburn Town fossil reef, 127; at Coral Gardens reef, 159; at Devil's Point fossil reef, 123, 124, 125; in Florida Keys reef, 86
boring sponges, 86, 125–26
boulder star coral, 91, 148–49
brain coral, 1, 12, 13, 16, 18, 123, 180, 201
branching corals, 87, 108, 146
brine shrimp, 122
*Broken Nature* exhibit (MoMA), 188–89
bryozoans, in death assemblage, 85
burrows, 33, 124
buttresses, 60

calcium carbonate ($CaCO_3$), 15, 21, 85, 127–28
caliche, 123
carbon dioxide emissions, 92, 118–19, 222, 224, 225
Caribbean/Caribbean reefs: *Acropora* corals in, 148, 171; coral nurseries in, 171–72; as cul-de-sac of the ocean, 70; currents in, 161; decline of, 69–70, 93, 97–98; disease in, 96; disturbances to, 116; elkhorn coral in, 93, 106, 206; fossil reefs in, 20; human changing to, 101; Hurricane Allen in, 96; isolation of, 70; overfishing and, 108–16; parrotfish in, 112–15; persistence of, 97; pillar coral in, 215; recovery chances of, 116; refugia search in, 161; resilience of, 97; sea temperatures in, 158–59; sea urchins in, 95; species in, 71; staghorn coral in, 93, 106, 206; statistics regarding, 93; stony coral tissue loss disease (SCTLD) in, 180–81; threats to coral in, 203; timeline regarding, 105–6; water quality in, 106–8; white band disease in, 89–90, 94–95, 150, 221. *See also specific locations*
Carne, Lisa, 174, 176, 178, 179–80, 184, 201, 202, 234
Carysfort Reef, 76–77
Carysfort Reef Light, 77
cauliflower coral, 193, 194–95, 207, 212–13
Cenozoic era, 5

Channel Caye, Belize, 89–90, 95
Chile, earthquake in, 43
Christ of the Abyss, 85
cirrus clouds, 190
clams: encrusting death assemblage, 85, 86; in fossil reefs, 20, 55, 125–26
climate change: actions regarding, 224; control of, 231–32; effects of, on coral reefs, 3, 94, 119; in El Niño-Southern Oscillation (ENSO), 220; future of, 119, 162–63, 231–32; as greatest threat to reef health, 160; greenhouse gas emission reduction for, 162, 190; intensification of, 157; limestone production and, 223; limiting, 162, 189–90, 224; mangroves and, 152; projections and likely coral survival, 223; refugia from, 141; sea levels and, 131; sea temperature change and, 159; sediment runoff in, 183; strategies for, 190; water temperature and, 92, 118. *See also* ice age
climax community, forming of, 125
cloud condensation nuclei (CCN), 191
clouds, 190–91
coal mining, 196–97
coastal development, risks in, 152, 231
Cockburn Town, San Salvador, 55
Cockburn Town fossil reef (San Salvador): coralclams at, 126–27; coral species pattern in, 58; destruction to, 126–27; ecosystem of, 62–63; erosion in, 63, 123, 127; introduction to, 56; pattern of reef zones in, 63; photo of, 128; sea level changes at, 127, 129, 130, 231; unconformity at, 127
coconut trees, 49
Cocos Keeling Islands, 42, 45–46, 47, 52, 66
Cocos Malay, 46–47
Cohen, Anne, 211
Cold War, 52
Connell, Joseph, 116
coral bleaching: adaptation to prevent, 195; causes of, 118; from climate change,

69; clouding and, 190–91; conditions causing, 198; at Coral Gardens Reef (Belize), 219; in coral nurseries, 193; defined, 3, 19; degree heating weeks (DHW) and, 198–200; effects on fish, 220; effects on invertebrates, 220; in El Niño-Southern Oscillation (ENSO), 219–20; expansion of, 94; experiment regarding, 194–95, 200; global events, 219; in Golfo Dulce (Costa Rica), 193; in Great Barrier Reef (Australia), 99; hunting capabilities during, 192; increase of, 160; internal waves and, 163–64; marine cloud brightening for, 191; overview of, 19; persistence during, 200; recovery from, 213; seawater mist for preventing, 190; in Thailand, 163; in turbid waters, 166–67; water temperature and, 92. *See also* heat tolerance

Coral City Camera, 2

coralclams, 126

coral disease, 3, 160. *See also specific diseases*

Coral Gardens Reef (Belize): *Acropora prolifera* in, 148–49; boulder star coral in, 148–49; coral bleaching at, 219; description of, 155; destruction of, 190; elkhorn coral in, 148–49; fish in, 155–56; photo of, 155; refugia in, 148–49, 219; sea temperature change in, 157, 159; staghorn corals in, 148–50, 154–55, 156–57, 158, 219; topography of, 155

*Coral Ghosts* (documentary), 69

coral larvae, 140, 186–87, 188

coralline algae, 21, 85, 87

corallites: of elkhorn coral, 63; growth of, 16–18; of staghorn coral, 167

coral nursery: coral bleaching in, 193; coral types in, 171; description of, 169; development of, 171–72; fragmentation in, 171; in Golfo Dulce (Costa Rica), 183; growth in, 169; at Hol Chan Marine Reserve (Belize), 168–69; locations for, 173; stony coral tissue loss disease (SCTLD) in, 170; structure of, 169, 172, 182

coral reefs: age of, 5; benefits of, 229; coastal development and, 152; concerns regarding, 3, 8; death assemblage in, 77; decline of, 69–70; future for, 8–9, 28; geological shutdown of, 137; glacier impact on, 129; hiatus in, 134; historical contact with, 102; human populations near, 151; life assemblage in, 77; location of, 5, 26; management process of, 179; multidisciplinary effort regarding, 11; paradox of, 217–18; pattern of zones of, 58–59, 60, 61–62; persistence of, 9–10, 72; phase shifts in, 97, 99; resilience of, 9–10, 71; shark populations and, 111; shifting baselines in, 100–101; soundtrack in, 187; statistics regarding, 229; types of, 25–26; uniqueness of, 61; wave energy in, 61, 62; weakening of, 3. *See also* fossil coral reefs; specific reefs

*Coral Reefs and Atolls* (Gardiner), 51

Coral Reef Watch (NOAA), 159

corals: acclimatization of, 204, 209–10; adaptation of, 195, 197–200; calculation of, 156; changes to, 3–4; as collectives, 205; cyclical changes and, 3–4; disturbances and, 16; environments of, 19, 135; estimation of percent live, 156; extinction of, 70; feeding, 191–92; fossilization chances of, 80–81; fragmentation of, 171; genetic diversity of, 176; growth process of, 42; microbiomes of, 205; number of species, 70–71; origin of, 129; polyps of, 2–3, 16, 17, 18, 19, 205; power of, 3; relocation of, 140; reproduction of, 18–19, 171, 178, 202; resilience of, 144; as sessile, 129; shape of, 60–61; spaces between, 20; sunlight importance to, 132, 165; at sunlit depths, 44, 48; tentacles of, 192; threats to, 210; upward growth of, 45; and wave energy, 48–49, 61. *See also specific types*

Coral Triangle, internal waves in, 165

Covid lockdown, 140, 141

Cramer, Katie, 105–8, 113, 115, 119, 174, 183

crocodiles, 65–66

crown-of-thorns starfish, 99
crust, Earth's, 130
Crutzen, Paul J., 7
cumulus clouds, 190
Curaçao, 66–67
Curran, Al, 57, 121, 123, 124, 144, 149, 151, 153, 215, 216
Curran, Jane, 153, 215
currents, sea temperature and, 161

Dairy Bull Reef, 98
damselfish, 3, 187
Darwin, Charles: on atolls, 26, 43, 44, 45, 47; in Chile, 43; at Cocos Keeling Islands, 42, 45–46; on the Cocos Malays, 46–47; on coral growth, 42, 45; coral reef paradox of, 217; on coral zones, 48–49; on criticism, 50–51; influences, 40–41; on Isla Santa María, 43; *On the Origin of Species*, 196; pigeon interest of, 195; *Principles of Geology* (Lyell) and, 40–41; *The Structure and Distribution of Coral Reefs*, 45–46, 50; on subsidence, 43; in Tahiti, 44–45; theory sharing by, 50; uniformitarianism and, 36, 41; on uplift, 52; *The Voyage of the Beagle*, 41
date mussels, 126
death assemblage, 77, 78–80, 84, 85, 86, 87–88. *See also* skeletons
debris, for artificial reefs, 185–86
deep reef refugia hypothesis, 165–66
degree heating weeks (DHW), 198–200
Devil's Point fossil reef: conditions of, 121; coralclams at, 126; coral return in, 219; corals at, 128–29; date mussels at, 126; description of, 122, 124; erosion at, 124; reef identification in, 123; sea level changes at, 123, 124, 126, 129, 130; staghorn coral at, 128; unconformity at, 123, 124, 125, 126, 128, 228
Discovery Bay, Jamaica, 97, 98
disturbances, 116–18, 122, 206. *See also* climate change; environmental changes; sea levels; sea temperature
Dobbs, David, 51

Dominican Republic, 89, 149
drowned reef, 132–33, 139
Dry Tortugas National Park, 76

Earth, 5, 26, 36, 130
earthquakes, 43
Eastern Tropical Pacific: internal waves in, 165; marginal reefs in, 135; as paltry, 213; reef challenges in, 193; reef return in, 139; sea temperatures in, 158–59; unconformity at, 145. *See also specific locations*
ecological succession, 125
ecosystems, 71, 97, 116–17, 125
Efremov, Ivan, 83
Egypt, 67–68
elkhorn coral: ages of, 171; in Caribbean reefs, 93, 106, 206; in Cockburn Town fossil reef, 62–63; in Coral Gardens Reef (Belize), 148–49; corallites of, 63; decline of, 106; description of, 20, 59, 63; disease of, 68, 71; environment for, 59; in Hurricane Allen, 96; in Jamaica, 62; in Laughing Bird Caye (Belize), 175, 176, 178; observation of, 62; photo of, 64; in reef crests, 59, 63, 67, 71, 72; reproduction of, 178; restoration projects for, 181; in sea level changes, 148; sugar plantations and, 108; in Telephone Pole Reef, 73; in Treasure Beach, Jamaica, 130; white band disease and, 94–95, 96–97
El Niño-Southern Oscillation (ENSO), 3, 135–36, 157, 159, 219–20
encrusting corals, 97
encrusting organisms, 85, 125
endangerment, 203, 214–15, 219
Enewetak Atoll, 51–52
environmental changes: acclimatization to, 204; as cyclical, 222; examples of, 210, 221; extinction avoidance during, 139; gene expression and, 204; geological shutdown of reefs in, 137; natural selection and, 196–97; research regarding, 134–35; unconformity and, 134. *See also* climate change; disturbances

epigenetic inheritance, 204
equator, refugia and, 162–63
Eritrea, 68
erosion, 36–37, 63, 65, 133, 139, 147
evolution, 147, 196, 201–2. *See also* natural selection
extinction, 70, 133, 142, 219

Faux, Victor, 174, 179
Fernandez Bay (San Salvador, Bahamas), 72
fertilizers, 107
50 Reefs initiative, 160
finches, 147
Fine, Maoz, 198, 200
finger coral, 2, 59, 73
fish: bioerosion by, 86; coral bleaching effects on, 220; food for, 22; fossilization of, 83, 113–14; herbivores, 112–16; poop, in limestone, 22; population increase, 225; preservation of, 22; at shipwrecks, 185–86; as soundtrack, 187. *See also* overfishing; specific fish
FitzRoy, Robert, 42, 48
flamingos, 122
Florida Keys: charting of, 103; drowning reefs in, 139; environmental stressors in, 138; expansion and contraction of reefs in, 137; Florida Reef Tract in, 76, 85, 90–91, 92; fossil reefs in, 65; limestone loss in, 137; map comparison of, 104–5; overfishing in, 109; reef restoration in, 181–82; runoff and declining water quality in, 107; sea temperature in, 138; shallow reef zones in, 76; stony coral tissue loss disease (SCTLD) in, 220; uniformitarianism in, 137
Florida Keys National Marine Sanctuary, 76, 181
Florida Reef Tract, 76, 85, 90–91, 92
footprints (tracks), 33–34
foraminifera (forams), 21, 85
forams, 85
forereef, 60–61
Forman-Castillo, Kirah, 168–73, 181

formation (geological), defined, 57
Fort Lauderdale, Florida, 186
fossil coral reefs: disturbances in, 117; ecological record in, 14–15; ecological succession, 125–26; erosion of, 65; insight from, 4; lack of preservation of, 65; locations of, 4, 14, 26; pattern of reef zones in, 62–65; sea level records in, 129–30; on sinking atolls, 52; time averaging of, 125; uplift of, 66–68, 117. *See also* coral reefs; specific locations
fossil fuels, 222–23
fossilization, 80–81, 83, 85
fossil record: analysis of, 227–28; bias in, 82; challenge of finding resilience in, 97; gaps in, 228; incompleteness of, 83; persistence evidence in, 122; refugia in, 141–42; taphonomy and, 83; unconformity in, 229–30
fossil soil (paleosol), 57
fragmentation, process of, 171
Fragments of Hope, 173, 174–81, 201–2
fringing reefs, 26
Frost Museum of Science, 2, 9, 234
functional redundancy, 71

Galápagos Islands, 165
gametes, 19
Gardiner, John Stanley, 51
Garifuna, 115
gas (fossil fuel), 222–23
Gauld, George, 103–4
Geister, Jörn, 61
gene expression, 203–4, 209–10
genetics, 147, 171, 176, 195
geologic time scale, 5–7
geology, 15, 33
Gerace Research Centre, 32, 54
ghost shrimp, 33
glacial periods: challenges in, 142; defined, 26; ice sheets and glaciers in, 129; limestone platforms in, 143, 144, 145; reef refugia during, 145; sea level changes in, 131–32, 142; shallow tropical ocean during, 142; survival on land during, 141

glaciers, 129, 130, 138. *See also* ice
Goad, Alex, 188
Godfrey, Dale, 174
Golfo Dulce (Costa Rica), 183–84, 193, 194, 212
goliath groupers, 109–11
Goreau, Nora, 62, 96
Goreau, Thomas F., 62, 65, 96, 101, 106
Goreau, Thomas J., 68, 69
Grand Cayman, 65, 111–12
Great Bahama Bank, 28–29
Great Barrier Reef (Australia): *Acropora* corals in, 108; annual report on, 99–100; climate models for, 161–62; coral bleaching in, 99; crown-of-thorns starfish in, 99; currents in, 161; disturbances to, 116; limestone under, 26, 29; marine cloud brightening in, 190–91; phase shift in, 99; protected areas of, 225; refugia in, 147, 161, 162–63; sea level changes in, 132–33, 137; sea temperature at, 161, 162; threats to, 99; water quality changes in, 108, 166–67
Great Barrier Reef Foundation, 190
Great Inagua, 63, 122, 137, 144
great star coral, 1. *See also* star coral
greenhouse gas emissions, 162, 190, 222, 224–25
greenhouse periods, 26
Greenstein, Ben, 75, 90
Greer, Lisa: Belize research of, 149–50, 153, 155–59; Dominican Republic research of, 88–89, 149; on Hurricane Earl, 158; at Laughing Bird Caye (Belize), 201, 226, 233; photo of, 155; on pillar coral, 215, 216; on reef restoration, 168, 226; on staghorn coral, 158, 159
Grotto Beach Formation, 57
Grottoli, Andréa, 200
Gulf of Aden, 198
Gulf of Aqaba, 198, 200
Gulf of Mexico, 147–48

*Halimeda* algae, 20–21, 33
Hall, Sir James, 37

hardground surfaces, fossils on, 124–25
Harrison, Daniel, 190
Hawaii, 132, 142, 192, 226
heat tolerance: of brain coral, 201; of cauliflower coral, 207; in coral nurseries, 194–95; degree heating weeks (DHW) and, 198–200; experiment regarding, 200, 209–10; genetics for, 201; in Gulf of Aqaba, 198; human-assisted evolution for, 201; in Laughing Bird Caye National Park (Belize), 201–2; map of Red Sea coral adapted for, 199; microbiomes and, 206; observation of, 201–2; ocean modeling to find locations of, 211; in Ofu (American Samoa), 210; at the Red Sea, 197–98; reproduction for, 202; symbionts for, 207; of tabletop coral, 208–9; testing of, 211; time for adaptation for, 200–201. *See also* coral bleaching; sea temperature
Hell (Grand Cayman), rock features at, 65
hermit crab, 34
historical ecology, 102
Hol Chan Marine Reserve (Belize), 168–69, 170, 172, 173, 177, 216
Holland, Steve, 33–34
Holocene thermal maximum, 138
Holocene time, 5, 26
*Homotrema rubrum*, 85
Hooker, Joseph, 196
human-assisted evolution, 201
humanity, as paradox, 232
Huon Peninsula, Papua New Guinea, 67
Hurricane Allen, 68, 96
Hurricane Earl, 157–58
hurricanes, 68, 94, 96, 119, 157–58
Hutton, James, 36–40, 42–43, 123, 228
hybrid reefs, 189

ice, 129, 130, 138. *See also* glaciers
ice ages, 26, 70, 129, 141
*Illustrations of the Huttonian Theory of the Earth* (Playfair), 40
imperialism, 41
Inagua National Park, 122

Inagua parrots, 122
Indian Ocean, 67, 142, 198, 201
individuals, in natural selection, 204–5
Indonesia, 29, 189
Indo-Pacific/Indo-Pacific reefs, 61, 70, 71, 99
interglacial periods, 26, 129, 141, 143, 144, 148
Intergovernmental Panel on Climate Change (IPCC), 119, 160
internal waves, 163–65
International Commission on Stratigraphy, 7
International Ocean Discovery Program (IODP; formerly known as the Deep Sea Drilling Project), 132, 133
International Union for Conservation of Nature (IUCN), 93
invertebrates, coral bleaching effects on, 220
Isla Santa María, Chile, 43, 52
Isthmus of Panama, 70

Jackson, Jeremy, 72, 101, 105
Jamaica, 63, 65, 68, 69, 96
James Hutton Trail, Scotland, 37
John Pennekamp Coral Reef State Park, 76
Joulter Cays, 21

Kelly, Mark, 204
Kelly, Scott, 204
Key West, Florida, overfishing in, 109. *See also* Florida Keys
Kleypas, Joanie, 164, 183–84, 193, 195, 212, 213
Knowlton, Nancy, 225
Konzhukova, Elena Dometevna, 83

Lagerstätten, forming of, 82
Lake Rosa, Great Inagua, 122
La Niña, 3, 136
Laughing Bird Caye National Park (Belize), 174–81, 201–2, 232–33
lava, 25
*Lazarus taxa*, 142
life assemblage, 77, 79, 87
lime mud, 22
limestone: abundance of, 23; accounting of, 136; atolls and, 44–45; chemistry of, 15; climate change and, 223; destruction of, 136; dissolving of, 123; forming of, 15–16, 19, 21, 27–28, 29, 143; fossils in, 20; in glacial periods, 143, 144, 145; in interglacial periods, 143, 144; layers of, 23–24; locations for, 23; oil and gas in, 222–23; platforms of, 24–25, 27–29, 143; quarries of, 1, 56; reef geologic growth of, 136, 137; reef restoration and production of, 181; in San Salvador, 14; shells in, 15; sinking of, 24–25; skeletons in, 15. *See also specific locations*
Line Islands, 230
lobsters, 3
Lucayan communities, 102
Lyell, Mary, 40, 50
Lyell, Sir Charles, 40, 49, 50, 196

macroalgae: in Caribbean reefs, 93–94, 98; on coral skeletons, 173; defined, 93–94; in Discovery Bay, 97; on finger coral, 73; grazing of, 116; green turtles and, 111; growth of, 69, 96; management of, 181–82; parrotfish and, 112; phase shift and, 97–98; proportion challenges of, 98; sea urchins and, 95–96; statistics regarding, 98; water quality and, 107–8
Malaysia, 29
Maldives, 29, 188
mangroves, 152, 225
marginal reefs, 135, 136, 138, 183, 193
marine cloud brightening, process of, 191
marine hardgrounds, 124–25
marine heat waves, 219. *See also* sea temperature
Marshall Islands, 211
Martin, Anthony, 22–23
massive corals, 62, 97
Maurice A. Ferré Park, 1
Maya communities, 102
Maya hamlet, 179
McClenachan, Loren, 104, 109, 119
McWhorter, Jennifer, 161–62, 166
medicine, coral reef benefits to, 229

Mesoamerican Barrier Reef, 26, 89, 102, 151, 165
Miami Beach, Florida, 189
microbes/microbiomes, 3, 21, 205–7
microfragmentation, 180, 181
mineral crystals, 127–28
Mission: Iconic Reefs (NOAA), 181, 182
MIT General Circulation Model, 164–65
Modular Artificial Reef Structure (MARS), 188–89
mollusks, 55–56
Monroe County Public Library (Florida), 109
Mooney, Aran, 187
Mo'orea, 44
Moraga, José Andrés Marin, 193, 195
Moreton Bay, Australia, 108
Morton Salt, 122
mudstone, forming of, 38
Museum of Modern Art (New York City), 188–89

National Oceanic and Atmospheric Administration (NOAA), 75, 159, 161, 181
natural selection, 195–97, 200–201, 204–5, 207. *See also* adaptation; evolution
The Nature Conservancy, 211
nautical charts, 103
Neil, Amir, 174
Nixon, Henry, 122–23
Norris, Dick, 113
North Keeling Island, 47
North Point, San Salvador, 34–35
nuclear testing, 51–52
nurse shark, 169. *See also* sharks

OceanOptimism, 225–26
oceans: acidity of, 118–19; carbon dioxide in, 118–19; circulation and the Caribbean, 70; modeling of, 164–65, 211; movement of, 70; renewable energy from, 225; shifting baseline syndrome in, 100–101; temperature rise of, 71, 92, 118

Ofu (American Samoa), 207–9, 210, 229–30
oil (fossil fuel), 222–23
Oliver, Thomas, 209
Olson, Randy, 96
*On the Origin of Species* (Darwin), 196
ooids, 21–22
opportunistic coral, 97
*Orbicella. See* star coral
organ pipe orbicella (*Orbicella nancyi*), 133–34
overfishing, 69, 101, 108–16, 225
Owl's Hole Formation, 57

Pacific Ocean, 135, 142, 148
Palau, 211
paleoecology, 57
paleontology, 15
paleosol (fossil soil), 57
Palm Islands (Australia), 88
Palumbi, Anthony, 210
Palumbi, Steve, 208, 209, 210, 211, 218, 226
Panama, 135–36, 137, 207
Pandolfi, John, 67, 72, 75, 117–18, 134, 227
Papua New Guinea, 29, 117–18, 218–19
Paris Agreement, 223–24
parrotfish, 22–23, 33, 112–13, 114–15, 187, 225
parrots, 122
patchiness, benefits of, 116
patch reefs, 59, 86
Patterson, Joshua, 182
Pauly, Daniel, 100–101
Peckol, Paulette, 149
peppered moths, 196–97
persistence, 97, 124, 128, 133, 150, 200
*The Peterson Field Guide to Coral Reefs*, 58
phase shifts, 97, 99
photosynthesis, 19, 61, 166, 191–92, 206
phytoplankton, 108
pigeons, 195
pillar coral, 169–70, 214–16, 217, 232–33, 234
plankton, 187
Plant a Million Corals Foundation, 180

plate corals, 62
Playfair, John, 37, 38–40, 49
Pleistocene reefs, 56–57, 102, 106
Pleistocene time, 5
*Pocillopora*. *See* cauliflower coral
*Pocillopora palmata*, 134
pollution, water, 225
polyps, 2–3, 16, 17, 18, 19, 205
poop, fish, 22–23
*Porites porites*. *See* finger coral
Port of Miami, Florida, 1, 2, 220
post-traumatic stress disorder (PTSD), 9
Precht, Bill, 89, 95, 174, 221
present, as key to the past, 35, 36, 40, 74
*Principles of Geology* (Lyell), 40
prolifera coral, 171, 172, 176. See also *Acropora prolifera*
Proterozoic time, 5

quarries, for building stone, 56

Raising Coral, 183–84, 212
Red List of Threatened Species (IUCN), 93
Red Sea, 29, 67–68, 130, 142, 197–98, 199
reef cores, 113–14, 132–33, 134
reef crest, 59, 60, 67, 71, 72, 91
Reef Design Lab, 188
reef ecology, analysis of, 57–58
*Reef Madness* (Dobbs), 51
Reef Renewal USA, 182–83
reef restoration: benefits of, 234; defined, 184; effectiveness of, 226; in Florida Keys, 181–82; heat tolerance in, 195; in Indonesia, 189; process of, 201, 226; purpose of, 183; spiders structures in, 189; types of, 184. *See also* coral nursery
reef sharks, overfishing of, 110–11. *See also* sharks
refugia: characteristics of, 144, 147; in climate change, 141; Coral Gardens Reef (Belize) as, 148–49, 219; deep reef refugia hypothesis and, 165–66; defined, 141; equator and, 162–63; future of, 160; 231–32; genetic evidence of, 147; in glacial periods, 145; in Great Barrier Reef (Australia), 147, 161, 162–63; in Gulf of Aqaba, 200; in interglacial periods, 148; internal waves and, 163–64; as large scale, 141; scales of, 141; searching for, 160; size of, 141; as source of hope, 160, 230; turbid waters as, 166–67; in twilight zone, 166. *See also* reef restoration
renewable energy, 224–25
reproduction, coral, 19, 171, 178, 202
resilience: of Caribbean reefs, 97; of corals, 144; defined, 9, 71; differences in, 218; fossil record implications for, 97; limitations of, 100, 159; low, 122; of staghorn coral, 149; of tabletop coral, 208–9; of zooxanthellae, 206–7
Richter, Rudolf, 83
ripples, in rocks, 38
rocks, 15, 37, 38
Rothko, Mark, 23
Royal Society of Edinburgh, 37
runoff, 106–8, 226

Sala, Enric, 230
sand dunes, 34–35
San Pedro, Belize, 151, 152–53, 157–58
San Salvador: beachrock at, 35; description of, 13–14; driving in, 54–55; fossil reefs in, 55; geography of, 14; Grotto Beach Formation in, 57; limestone in, 19–20, 21, 23, 25, 28, 29; Owl's Hole Formation in, 57; research of, 32–33; sea level changes at, 144; uniformitarianism in, 33, 35
sargassum, 93–94, 152–53
scallops, 3
Scotland, 36–40
scuba diving, 78, 96, 100–101, 145–46, 149, 154–55, 185
sea anemones, 16
sea fans, 146
seafood, evidence of ancestral eating of, 68
seagrass, 111

sea levels: climate change and, 131; at Cockburn Town fossil reef, 127, 129, 130, 231; coral reef formation and, 66; corals and, 131–32; at Devil's Point, 123, 124, 126, 129, 130; drowned reefs and, 132; Earth's crust and, 130; extinction and, 133; in fossil coral reefs, 129–30; future effects of, 231; in glacial periods, 131–32, 142; at Great Barrier Reef (Australia), 130, 132–33, 137; in Great Inagua, 137, 144; ice age cycles and, 52; rapid changes to, 131–32; at the Red Sea, 130, 142; reef disruption by, 27; at San Salvador, 144; at Treasure Beach, Jamaica, 130; vulnerabilities regarding, 131; at Yucatán fossil reefs, 130

sea temperature: in the Caribbean, 158–59; coral adaptation for, 195; at Coral Gardens Reef (Belize), 157, 159; currents and, 161; degree heating weeks (DHW) and, 198–200; in Eastern Tropical Pacific, 158–59; effects of suspended sediment on, 166; El Niño and, 159; evidence preserved in coral skeletons, 133; in Florida Keys, 138; in Golfo Dulce (Costa Rica), 212; at Great Barrier Reef (Australia), 161, 162; National Oceanic and Atmospheric Administration (NOAA) and, 161; in Ofu (American Samoa), 208; sensors for recording, 157; variations of, 163. *See also* heat tolerance

sea turtles, 3, 111–12, 225–26
sea urchins, 69, 86, 94, 95–96, 98, 181–82
seawalls, development of, 151
seawater, building blocks from, 15
seawater mist, 190
sea whips, 146
sediment: burial for fossilization, 81; dangers to coral, 106, 166, 183; erosion creating, 36–37; on limestone platforms, 143; of parrotfish, 22–23; runoff increasing as climate changes, 183; sea temperature and, 166
sedimentary rocks, 34–35, 36–40, 127
shallow reef zones, 76, 85

sharks, 3, 78, 110–11, 147–48, 169
shells, 15, 20, 72, 81–82
shifting baseline syndrome, 100–101
shipwrecks, as artificial reefs, 185–86
shrimp, 3, 147, 187
Siccar Point, Scotland, 37–40, 123–24
sinking, support for, 49
*Sixth Assessment Report* (IPCC), 162
skeletons: burying of, 81; calcium carbonate ($CaCO_3$) and, 15; coral bleaching and, 19; of corals, 16, 18; degradation of, 88; destruction of, 82; examination of, 84; fossilization of, 80–81; growth of, 18; at Hol Chan Marine Reserve (Belize), 172; illustration of, 18; in limestone, 15; macroalgae on, 173; photo of, 17; on pillar coral, 216; research regarding, 75–76; in Telephone Pole Reef, 73–74
Smithsonian National Museum of Natural History, 80
Smithsonian Tropical Research Institute, 113
snails, 3, 20
snappers, 110
soft corals, 146
solar energy, 224–25
soldierfish, 187
sound, for attracting coral larvae, 187
sounding line, process of, 103
South Keeling Islands, 47
spiders, in reef restoration structures, 189
staghorn coral: ages of, 171; in backreefs, 59; calculation of, 156–57; in Caribbean reefs, 93, 106, 206; in Channel Caye, Belize, 89–90; in Coral Gardens Reef (Belize), 148–50, 154–55, 156–57, 158, 219; in Dairy Bull Reef, 98; in death assemblage, 87–88; death of, 88–89; description of, 20, 154–55; at Devil's Point, 128; disease of, 68; in the Dominican Republic, 89, 149; excavation of, 149–50; fish in, 156; in Florida Reef Tract, 90–91; forming of, 149; in Frost Museum of Science, 2; growth of, 173; at Hol Chan Marine Reserve (Belize), 172,

173; in Hurricane Allen, 96; at Laughing Bird Caye (Belize), 175; in the Mesoamerican Barrier Reef, 89, 165; observation of, 62; persistence of, 150; recovery of, 98; resilience of, 149; restoration projects for, 172, 173, 180, 181; in sea level changes, 148; study of, 89; in Telephone Pole Reef, 73; variations of, 176; white band disease and, 89–90, 94–95, 96
star coral, 1, 16, 17, 59, 180
stone tools, discovery of, 68
stony coral tissue loss disease (SCTLD), 92, 170, 180–81, 216, 220, 221
*The Structure and Distribution of Coral Reefs* (Darwin), 45–46, 50
subfossils, defined, 89
subsidence, 43, 49
Sue Point, San Salvador, 55
Summer Island (Maldives), 188
sunlight, 132, 165, 166, 190–91
Super Reefs Initiative, 210–11
survival of the fittest, 195

tabletop coral, 208–9
Tahiti, 44–45, 132
taphonomy, 83
Tarracino, Coral, 110
Tarracino, Mae, 110
Tarracino, Tony, 110
tectonic plates, 27, 51, 66
Tela Bay (Honduras), 230
Telephone Pole Reef, 73–74, 88, 95, 102
tentacles, coral, 192
Thailand, 163–64
3D printing, for artificial reefs, 188
tides, internal waves and, 163
time averaging, 125
Tortuguero, Costa Rica, 111–12
Toth, Lauren, 134–35, 180, 181, 231
trace fossil, 86
Treasure Beach, Jamaica, 130
trees, ecological succession and, 125
trilobite, 80
turbid waters, as refugia, 166–67

turtles, 3, 111, 225–26
twilight zone, 166

unconformity: at Cockburn Town fossil reef, 127; defined, 38–39; at Devil's Point, 123, 124, 125, 126, 128, 228; at Eastern Tropical Pacific, 145; environmental changes and, 134; in fossil record, 229–30; future layer of, 229–30; future view of, 229; at Siccar Point, Scotland, 123–24
uniformitarianism: Charles Darwin and, 36, 41; defined, 33; earthquakes and, 43; in Florida Keys, 137; James Hutton and, 36; limitations of, 35, 74; as puzzle, 42–43; rejection of, 36; in San Salvador, 33, 35; of sedimentary rocks, 34–35; support for, 40
United Nations Emissions Gap Report, 224
University of Miami, 189
uplift, 43, 52, 66
uranium-thorium dating, 88
Ussher, James, 36
USS *Oriskany*, 185
USS *Spiegel Grove*, 185
US Virgin Islands, 86, 187–88

Valdez, Narciso, 154, 169, 216
*Vandenberg*, 185
Vaughan, David, 180
Villalobos, Tatiana, 195
volcanoes/volcanic islands, 25, 45, 51, 117
*The Voyage of the Beagle* (Darwin), 41, 46

Walker, Sally, 29–30, 33–34, 121
Wallace, Alfred Russel, 196
wall dive, 145–47
water quality: decline in Caribbean, 106–8; defined, 106; environmental change to, 222; in evidence from fossil reefs, 133; pollution and, 225, 226; refugia and, 166–67; runoff and, 108
wave energy, 61, 62, 67
weedy coral, 97–98

White, Brian, 57, 124
white band disease, 89–90, 92, 94–95, 96
"Will you become a fossil?" game, 80
Wilmot, Inilek, 69
Wilson, Mark, 121, 123, 124, 126–27
wind energy, 224–25
Windward Passage, 63
Wirth, Karl, 153, 201, 214, 215, 233
Woods Hole Oceanographic Institution, 211

World Wildlife Fund Mesoamerica, 178
worms, 125

Xcaret theme park (Cancún, Mexico), 65

Yucatán fossil reefs, 65, 130

zooplankton, 16, 191–92, 200
zooxanthellae, 19, 165, 191–92, 198, 206–7